God after Einstein

God after Einstein

What's Really Going On in the Universe?

John F. Haught

Yale
UNIVERSITY PRESS
New Haven & London

Published with assistance from the Mary Cady Tew Memorial Fund.

Copyright © 2022 by Yale University.
All rights reserved.
This book may not be reproduced, in whole or in part, including illustrations, in any form (beyond that copying permitted by Sections 107 and 108 of the U.S. Copyright Law and except by reviewers for the public press), without written permission from the publishers.

Yale University Press books may be purchased in quantity for educational, business, or promotional use. For information, please e-mail sales.press@yale.edu (U.S. office) or sales@yaleup.co.uk (U.K. office).

Set in Bulmer type by Newgen North America, Austin, Texas.
Printed in the United States of America.

Library of Congress Control Number: 2021944045
ISBN 978-0-300-25119-7 (hardcover : alk. paper)

A catalogue record for this book is available from the British Library.

This paper meets the requirements of ANSI/NISO z39.48-1992 (Permanence of Paper).

10 9 8 7 6 5 4 3 2 1

To Evelyn

Contents

Acknowledgments ix

INTRODUCTION 1

ONE
God 6

TWO
Eternity 18

THREE
Time 34

FOUR
Mystery 49

FIVE
Meaning 63

SIX
Origins 80

SEVEN
Life 95

EIGHT
Thought 113

NINE
Freedom 129

TEN
Faith 144

ELEVEN
Hope 155

TWELVE
Compassion 171

THIRTEEN
Caring for Nature 184

CONCLUSION 196

Notes 203
Index 225

Acknowledgments

I WANT TO THANK MY FRIEND and colleague Charles A. O'Connor III for reading early versions of the manuscript and, as always, for making many helpful comments. I am indebted also to my wife, Evelyn, and my sister Elaine for their indispensable advice. Thanks to Jennifer Banks and Abbie Storch at Yale University Press for their skillful management of the publication process. Finally, special thanks to Mary Pasti for her careful corrections and exceptionally fine editing talents. It has been a great pleasure working with her in completing this book.

God after Einstein

Introduction

WHEN WE LOOK AT AN OBJECT, it has a background, and the background makes a difference in how we see the object. For example, when we study a painting in a museum, it has a solemnity it would not have if it were hanging in a basement or a garage. Our awareness of the background is only tacit, but it makes a difference in how we see the object.

Throughout most of Western history the background of human life, thought, and worship has been a stationary universe. From one age to the next people lived, mated, worked, suffered, prayed, and died against a cosmic background perceived as motionless. The heavens seemed to rotate in a perfect circle around the earth, and planets wandered around a bit, but the celestial canopy stayed mostly the same while the earth loitered passively in its assigned place. When people thought about meaning, virtue, suffering, death, and God, they took for granted the dependable profile of a fixed universe.

But four centuries ago, thanks to science, the old frame began to splinter, and the cosmos started to stir. Over the past two hundred years the formerly motionless universe burst into the foreground and began to command our focal attention. The earth turned out to have a history, and life sprang from a fount too deep in time to fathom. During the past century, again thanks to science, a previously unnoticed 13.8 billion-year-old cosmic story spilled forth, carrying everything along with it.

What does the new scientific cosmic story mean for our understanding of God? The present book is a response to this question. Theological awareness has for centuries been woven into pictures of an immobile universe. Who can deny that even today the sensibilities of most religious people, and not just Christians, are most at home with a prescientific understanding of the natural world? Theology has moved only slowly and often reluctantly away from the ancient images of a static universe in which our faith-traditions came to birth. Even though many Christians, including popes, have conceded the correctness of evolutionary biology and scientific cosmology, our spiritual lives remain fastened to an ideal of timelessness stamped into our souls long before the coming of modern science. I believe the religious instincts of most people, including the scientifically educated, remain mostly out of sync with our new narrative of the natural world.

No doubt, scientific ideas have had at least some influence on modern theology. After Galileo and Newton, for example, theologians began thinking of God along the lines of a supernatural mechanic who physically determines the course of cosmic events for all eternity. Evolutionary biology, however, made it increasingly difficult to detect signs of special divine creativity and providential design amid all the meanderings of life, evolution, and human history. So, in its search for a more secure setting, theology went looking for God in the scientifically inaccessible interior lives of human subjects. Christian spirituality still remains focused less on what is going on in the universe than on the hidden drama of working out one's personal salvation in the presence of a deity disconnected from time and history. Even today the lure of timelessness allows the faith of millions to remain otherworldly and privatized, untouched by the long passage of cosmic time.

With few exceptions, the full reality of time has yet to be taken seriously by experts in spirituality. A feeling for deep cosmic time is also virtually absent from academic theology and suburban homilies. In this book, however, I seek an understanding of God commensurate with the new understanding of nature and time that recent cosmology—the scientific study of the universe as a whole—has introduced into the world of thought. After Einstein's revelations, a fundamental question for theology is What is *really* going on in the universe?

I love the natural sciences and have never thought of them as contradicting my religious beliefs. My awareness of scientific discoveries, however, has made a difference in how I have come to think about God. What, then, do I mean when I recite the Nicene Creed with my fellow Christians today? Can I hold on to an ancient, prescientific confession of faith while fully embracing contemporary scientific understanding? How, for example, can I reconcile my belief in God with evolutionary biology, especially since the latter has led so many other science-lovers to atheism?[1]

I have addressed the question of evolution and faith in previous books, especially *God after Darwin*.[2] Here, though, I want to widen my earlier inquiries by asking what God means after Albert Einstein (1879–1955). I have come to believe that theology may rightly address the troublesome questions about God and evolution only if it first takes into account the universe that science began unveiling in the early twentieth century. The story of life, we can now see, is part of an immensely longer and more nuanced history of nature than Charles Darwin (1809–1882) and his early followers knew about. The question of what's going on in the story of life is now subsidiary to the question of what's going on in the story of the universe.

Most scientists and theologians have so far failed to look at evolutionary biology in terms of its wider cosmic setting. Yet both the journey of life and the religious pilgrimages of humans look new once we tie them back into the larger story of the physical universe. While Einstein was developing ideas on relativity, he was not yet aware that the cosmos he was sketching mathematically is undergoing a long birth-process. He needed the help of other scientists, such as Georges Lemaître (1894–1966), Edwin Hubble (1889–1955), and George Gamow (1904–1968), to start laying out in more detail the new story to which his calculations were pointing. Nevertheless, since it is especially his science of relativity that forms the underpinnings of our new story of the universe, I am titling this book *God after Einstein*.[3]

The following chapters differ from previous reflections on Einstein and God.[4] Most of the literature on this topic so far has focused on three questions arising from Einstein's scientific work. First, does Big Bang cosmology entail the existence of God? It was Einstein's general theory of relativity that laid out the main theoretical foundations of the view that the universe began with a Big Bang. Does this theory give scientific plausibility to biblical accounts of an initial creation of the world by God?[5] A second question, one that I find more distracting than promising, arises from Einstein's notion of time-dilation. Relativity predicts that time slows down or speeds up relative to the curvature of gravitational fields. Since relativity allows for time in one gravitational setting to pass faster or slower than in others, can we now fall back on the elasticity of time's passage to line up the biblical week of creation in the book of Genesis with the immensely longer sequence of stages depicted in contemporary scientific cosmology? Does relativity allow in principle that billions of years to us humans may be only "one day" for God, so that Genesis may now be read as scientifically factual?[6] Closely related to the second question is the third: What does the nonsimultaneity of special relativity—that there is no universally standard "now"—mean for how and when God gets to know what is going on in the cosmos?[7] A lot of theological speculation has surrounded these three questions, but in my opinion they are almost trivial when compared to other themes associated with Einstein that I will be visiting in these pages. More often than not, the three questions just mentioned, especially the second, presuppose questionable, often literalist, styles of biblical interpretation. Theology, I believe, can do better than this.

The point of theology, as I understand it, is to explore the "reasons for our hope" (1 Peter 3:15). After Einstein, theology cannot avoid asking whether the scientific discovery that the cosmos is a long story with interesting outcomes may in some way give us new reasons for hope—or perhaps just the opposite. In 1916, Einstein had not yet fully understood that his impressive theory of gravity was in principle overthrowing every ancient, medieval, and modern idea of an immobile universe. He did not yet realize that his general theory of relativity was bringing to light an immense and still-unfinished cosmic story now estimated to be around 13.8 billion years old.

Although Einstein did not intend to do so, he has given theology a whole new point of view from which to ask about meaning, truth, and God. So far, both contemporary intellectual culture and traditional Christian theology have found little, if any, significance, much less reasons for hope, in the long journey that nature has been on in its temporal coming-to-be. In these pages, however, I shall be arguing—with one eye on Einstein's science and another on his personal opinions about religion—that what has been going on in the universe is nothing less than a great awakening and that this awakening, if we examine it carefully, comes with good reasons for hope. Not least among these reasons is that the universe has recently given rise to inquiring minds, such as the one with which you will be examining and criticizing these pages.

Although I am a lay Christian (Roman Catholic) theologian, I am writing not only for my fellow believers but also for anyone who wonders whether there are good reasons for hope and whether science after Einstein has anything to contribute to that nearly universal question. This book, then, is a contribution to the more general contemporary conversation on the relationship of science to religion. Even though my starting point is that of a Christian, my inquiry intersects significantly with questions that arise from many other quarters as well. I trust that my invitation to have Einstein accompany me at least part of the way in the following investigations will also draw the attention, if not always the consent, of scientists, philosophers, and other admirers of the great physicist.

ONE

God

> In their struggle for the ethical good, teachers of religion must have the stature to give up the doctrine of a personal God.
> —Albert Einstein

> The main source of the present day conflicts between the spheres of religion and science lies in the concept of a personal God.
> —Albert Einstein

> I believe in one God, the Father almighty, maker of heaven and earth, of all things, visible and invisible.
> —Nicene Creed

OUR UNIVERSE IS STILL COMING into being. The most important discovery scientists have made during the past two centuries, in my opinion, is that the cosmos is a story still being told. Research in geology, biology, chemistry, physics, and astrophysics shows that our universe has been on a long journey, one that apparently is far from over. But where is it going? Does the journey have a goal? What is really going on in the universe?

And what about God? What can ancient ideas of a divine creator possibly mean if the physical world has an unpredictable future? Given our new scientific understanding of a world in process, what do divine creativity, providence, and redemption mean? Do they have any meaning at all? Is it reasonable for scientists and those who accept scientific truths to believe in a personal God or "look forward to . . . the life of the world to come"?[1] Above all, what does it mean for our understanding of God that the universe is still aborning and that it has shown itself to be a drama of gradual awakening? And what should we be doing with our lives here and now if the universe is still a work in progress?

Albert Einstein had provocative thoughts not only about the universe but also about God and other topics of theological interest. My focus in this book is on the idea of God after Einstein, but since Einstein's science also changes our understanding of everything else, I want to engage with him on the related notions of eternity, time, mystery, meaning, cosmic origins, life, mind, freedom, faith, hope, and love—and even on what his science means for contemporary ecological concerns. I will also have occasion to comment on the problem of evil in Einstein's universe.

Early in the twentieth century the great physicist published his famous "field equations" dealing with the universe and the implications of gravity. His mathematics suggested that the universe is not the same from age to age. At first, Einstein did not believe what his own figures were telling him, but ever since 1916, when he published his general theory of relativity, cosmologists have demonstrated that nature is an ongoing epic, with new and surprising outcomes popping up and breaking out for billions of years. Until quite recently we could not have guessed that the universe has been on such a long, wild adventure, and we had no idea that in all probability it will undergo future changes far beyond the ones that have already come to pass. Surely other surprises await.

Aware now that the universe is still being born, science has given theology a fertile new framework for thinking about the meaning of everything, including faith in God. It matters not only to science but also to theology that the natural world, after giving rise to life and mind, still seems far from finished. Nature—the term comes from the Latin word for "birth"—is just barely emerging from the dark womb of its obscure past to an unpredictable future. The picture of nature as waking up from a long sleep calls for a new understanding of nature's God.

This book, I have already warned you, is a work of theology. Generally speaking, theology is the quest for reasons to hope in spite of the threats of meaninglessness, suffering, and death. Christian theology, which is my own point of departure, is systematic reflection on the meaning and truth of faith in the God whom Christians identify as the "Father" of Jesus of Nazareth. It is this God, not a faceless prime mover or efficient cause, that Christians officially believe to be both the creator of all things and the imperishable liberator of all finite beings from the fate of nothingness. In this book I want to ask what the God of Jesus means to us if we think in depth about the Big Bang universe whose general outlines sprang from the mind of Einstein.

So far, theologians have paid only halfhearted attention to the scientific news, arriving with Einstein, that the universe is still being born. To be sure, many of them agree that life has evolved and that the universe has had a long history, but a large majority have yet to reflect earnestly on what the new cosmic story means for our spiritual lives. Most Christian thought has settled on religious ideas that fit more comfortably into prescientific pictures of the world.

I am interested here in what the idea of God may mean to Christians after Einstein. Most of what I have to say in these pages, however, should be of interest to Jews and Muslims as well. For that matter, what I have to say here may mean something new to anyone who wonders about the religious meaning of the new scientific cosmic story. Instead of "no God," as scientific skeptics propose, we may now look into the universe after Einstein for a "new God." This God will still be one who creates, heals, and offers reasons for hope, as the Abrahamic traditions have taught. But theology after Einstein will now have to undergo an unprecedented self-revision by proportioning its sense of God to the vast cosmic horizons that science has recently brought into view.[2]

Scientifically educated people nowadays often suspect that the physical universe has outgrown all of our inherited images of a deity. For many sincere thinkers the ancestral portraits of God no longer evoke the sentiment of worship. To satisfy their native spiritual longings, they slip their souls inside the vast cosmos itself rather than bow down before the ancient gods. For them, nature is big enough.

But is it really? What if the universe is waking up to something infinitely larger and deeper than itself? In producing the human species, the cosmos has given birth to the kind of awakening we call thought, and thought has given rise to increasingly new questions, including What is the meaning of it all? After Einstein, theology, of all disciplines, cannot help being interested in the question of what is going on in the universe. Science has now given theology the opportunity to think fresh thoughts about religion and God. Einstein's universe also calls for new thoughts about eternity, time, meaning, and other themes of interest to the world's religious traditions. Other disciplines provide ways to think about these topics too, of course, and I have no wish to deny their importance. But science after Einstein has given us a whole new understanding of the universe. We now have the opportunity for a *cosmological* perspective on all religious phenomena. Here my focus is on the question of what Christian thoughts about God might look like if we connect them to the new scientific story of a still-awakening universe.

Religion and God

People who believe in God belong to the wide world of religion. We may understand religion, generally speaking, as the human awakening to *indestructible rightness*. As religions evolved, they increasingly featured a sense that beyond, within, and deeper than our immediate experience there exists something imperishably *right*. Religion is a basic confidence that beyond the wrongness of suffering, perishing, moral evil, and death lies an elusive but enduring rightness. Religion is trust—or faith—that this mysterious rightness is more persistent, more important, more real, and more healing than anything given in immediate experience. Awakening to indestructible rightness is an adventure that humans share with no other species of life.[3]

Religion first emerged on Earth, at least vaguely, with reflective self-consciousness and the capacity for symbolism. Religious aspiration was already stirring in the most elementary human capacity to distinguish rightness from wrongness. During the course of human history the sense of indestructible rightness became more and more explicit. Starting around 800 BCE, a few sensitive religious innovators, at separate places on our planet, almost simultaneously began distinguishing, more deliberately than ever before, between the wrongness of suffering, moral evil, and death, on the one hand, and an imperishable transcendent order of liberating rightness, on the other. The point of authentic human existence, their religions agreed, was to attune one's life and actions to an imperishable rightness—later identified in Christian theology as God and in Christian philosophy as infinite being, meaning, truth, goodness, and beauty.[4]

Here I am using the encompassing term "rightness" not in a narrowly moral sense but as a stand-in for the whole slate of incorruptible values that serve as human ideals. Our religious ancestors gave culturally specific names to the elusive rightness that was calling them to live truthfully, gratefully, humbly, peacefully, and compassionately. For example, Buddhism in India and later in China aspired to *right* wisdom, *right* action, and *right* mindfulness. Buddhists did not talk about God, but they had names such as *dharma* for indestructible rightness. In China, Laozi called it the *dao*. In India, religious sages referred to it as *brahman*. In Greek culture, Plato called it *the Good*, and Epicurus named it *logos*. In Israel, the prophets associated indestructible rightness (Yahweh) with justice (*tzedek, tzedekah, mishpat*). The right way to live, they preached, is "to do justice, and to love kindness, and to walk humbly with your God."[5] Later, Jesus of Nazareth discovered his own religious identity in the company of the great prophetic preachers of justice. For him the rightness sought by his ancestors was now arriving anew in the Kingdom of God. Through the power of God's "spirit" a new epoch of justice and compassion was now dawning, a reign of rightness that would have no end.

Jesus's and Einstein's Jewish ancestors had developed their sense of transcendent rightness from long traditions of storytelling: stories about the call of Abraham to move his family into a new future; the Exodus from Egypt under the leadership of Moses; the Hebrew people's struggle for liberation from slavery, genocide, and insignificance; the nation's pilgrimage to a land overflowing with new possibilities; the establishment of God's covenant; Israel's mission to be a light for all the nations; the sending out of God's spirit to renew the face of the earth; the powerful words of prophets urging their people to remain hospitable to strangers, to be faithful to their God, to work for justice, and never give up hope. By telling and retelling stories about these events and encouragements, a new ideal of rightness came to birth in Israel, one that is unique among the world's many religious traditions: Israelite religion came to associate indestructible rightness with the capacity to make and keep promises. In doing so, it opened up a new future for the people and gave to their religion a hope-filled interest in the dimension of what is not-yet. The motif of God's not-yet-ness will be of great interest to us throughout these pages.

In the Bible what is indestructible is God's promise, on which all of reality is believed to rest everlastingly. The universe is shaped not so much by a principle of fairness as by a limitless font of fidelity, care, mercy, and superabundant generosity. The "word" of God is a word of promise. God (Yahweh or Elohim) is the name the Israelites gave to the one who makes and keeps promises, who will always be faithful, and who expects unconditional trust. God is the liberator whose word opens up a new future even in the face of what seem to be dead ends. The sense of promise gives people reasons to hope. The "right" kind of religious life, accordingly, begins with trust that God's promises will be fulfilled. Christians, too, are committed to living within the horizon of promise and to keep seeking reasons for hope.[6] They associate their understanding of God, however, not only with the Israelite hope but also with the generous, just, faithful, promising, and liberating deity that Jesus of Nazareth referred to as his Father.

Christianity's belief in God is one instance of the perennial human longing to link our fragile existence to a rightness that is immune to perishing. Faith in God reflects a widely shared human conviction that only a connection to what is indestructible can relieve our anxiety about death and give permanent meaning to our lives. This is why the Abrahamic traditions have protected at all costs the notion of divine everlastingness. Yet Judaism, Christianity, and Islam all profess, in different ways, that their imperishable God has somehow intersected with the perishable world of time. For Christians, God has even "become flesh." The entrance of God into time is the main reason for their hope.

If God remained altogether outside time, then suffering, anxiety, and death would remain unhealed. But if God gets tied too closely to time and transience, will not rightness itself become subject to the threat of nonbeing? Can divine rightness be transcendent and indestructible, on the one hand, but also caring enough to merge intimately with time, on the other? Whatever its relation to time, the question of whether time is real is central to theological inquiry. If, as many scientists still assume, time is not real, then the Christian belief that God is "incarnate" in time cannot be of much consolation or consequence. But if time is real and God has come into time, as Christians believe, how can God be indestructible? By wrestling with Einstein's understanding of time and eternity I believe a deeper understanding of both God and time may emerge, one that gives a promising direction to our quest for reasons to hope.

Let us experiment, therefore, with the thought that the indestructible rightness to which faith points is healing and liberating, not because God is outside of time or saves us from time, but because—in a way that I shall gradually develop—God is somehow not-yet. It is because God is not-yet that there is always room for hope. It is because God is not-yet that the passage of time is not a threat but the carrying out of a promise. Thinking about God as not-yet may sound strange. For Christians, after all, "God is, was, and ever shall be." Everything is encircled presently by divine care. The world is in God's hands. Let us seek God's face. God is everywhere. God is both alpha and omega.

Yes, but after Einstein, God is less alpha than omega.[7] God is less "up above" and more "up ahead." If God were pure presence, there would be no room for a future to come or a promise to be fulfilled. If time is real and if the world is to have a future, divine presence must now in some way be restrained in order to make room for time and creation. Religious people usually want God to be an eternal presence, however, because time is terrifying. Popular piety has led people to look for ways of escaping the rushing waters of time and the relentless perishing of moments. Believers and theologians have preferred a God who is an "eternal now" rather than a "yet-to-come." In tune with ancient Platonic thought, their minds and imaginations have associated God with a haven of timelessness where permanence quietly calms their fears and saves them from the uncertainties of living in time and history.

Through the many centuries that have passed since the death of Jesus, Christians, more often than not, have tried to keep the indestructible rightness they associate with this man from getting tied too closely to the corrosiveness of time that made it possible for him to suffer and die. Their quest for timelessness has led them to locate God in an "eternal present," immune to temporal passage. Apart from a timeless God, everything is eventually lost, or so they have assumed.

Yet Christian theologians have never felt completely comfortable separating eternity so sharply from time. The doctrine of the Trinity—that God is not just Father but also Son and Holy Spirit—expresses the early Christian longing to preserve both the eternity and the temporality of God. According to the Nicene Creed, Jesus was murdered in time by crucifixion, but he remains everlastingly "consubstantial" (one in being) with "God the Father." Christians believe, then, that in the man Jesus indestructible rightness has come into time, but in such a way that the passage of time cannot vanquish it. This belief requires, however, a special understanding of the meaning of eternity and the meaning of time. In the following two chapters I will reflect on these meanings in conversation with Einstein.

Christians believe that in Jesus the wrongness of death and evil is defeated forever, not by God's abolishing time, but by God's making time internal to the divine life. In Christianity the image of a forsaken, condemned, and executed man is stamped forever onto its sense of indestructible rightness. When Christians claim that Jesus is the Son of God and that the Son has existed "before all ages," they are claiming that what is indestructibly right is self-emptying love. They trust that selfless love is what creates "the heavens and the earth." Instead of pulling the created world safely away from time, however, infinite love receives the temporal world, in all its ambiguity, into itself. In Christian understanding, God has conquered wrongness not by negating time but by gathering every moment of time's passage into the divine life—where nothing is lost.[8]

This is not how religious people have usually thought about indestructible rightness. As theologian John Macquarrie writes: "That God should come into history, that he should come in humility, helplessness and poverty—this contradicted everything . . . that people had believed about the gods. It was the end of the power of deities, the Marduks, the Jupiters . . . yes, and even of Yahweh, to the extent that he had been misconstrued on the same model."⁹ The belief that God internalizes what happens in irreversible time cannot be thought of as marginal to Christianity's understanding of indestructible rightness. The idea of such a vulnerable God, however, finds no comfortable place in the history of religious consciousness, including that of Jews, Muslims—and even most Christians.

The God of Christian faith is one who, instead of being a governor, dominator, or dictator, undergoes a *kenosis* (that is, a pouring out) of infinite divine love into the temporal world, as manifested in the self-surrender of Jesus on the cross. Christian theology, therefore, is obliged not only to acknowledge but also to highlight the divine kenosis, referred to by a recent pope as "a grand and mysterious truth for the human mind, which finds it inconceivable that suffering and death can express a love which gives itself and seeks nothing in return."[10] This kenotic image of God also transforms the whole notion of divine power. The theologian Jürgen Moltmann remarks that the God of Christian faith "is not recognized by his power and glory in the world and in the history of the world, but through his helplessness and his death on the scandal of the cross of Jesus." Moltmann adds that "the gods of the power and riches of the world and world history then belong on the other side of the cross, for it was in their name that Jesus was crucified. The God of freedom, the human God, no longer has godlike rulers as his political representatives."[11]

The kenosis of God is the ultimate reason why time is real and God is not-yet. Divine presence is humbly withheld to make room for time and hence the opportunity for something other than God to come into being. Ultimately it is because God is not-yet that there is room for time and the coming of a new future. When I talk about God in this book, then, I am thinking of the strange deity who identifies fully with the self-emptying, crucified Christ and, by extension, with the struggling and suffering of all of life—the God who embraces and conquers transience by bringing all of time, not to a finish, but to a fulfillment. While I want to take into account as far as possible the experience of other religious traditions, I am obliged to start with the scandalous Christian belief that God is vulnerable and defenseless love. I cannot casually pass over the fundamental Christian belief that God has chosen to be identified with a crucified man who was fully subject to the terrifying irreversibility of time.[12]

Christian theologians have found it hard to understand how divinity can be indestructible if God, having become incarnate in Christ, undergoes suffering and death in time. For if God becomes fully incarnate in time, how can God be indestructible? And how can divine indestructibility be right—that is, truly healing—if it fails to share fully in the passage of time and, along with it, life's struggling, suffering, and perishing?

The question of time—and hence the need to bring in Einstein's ideas—is of vital importance to theology as well as cosmology. Early Christians wrestled with the question of God and time by asking what it means to call Jesus the Son of God. How can indestructible goodness be identified with a crucified outlaw? they wondered. One way of resolving the paradox was to dissolve time into eternity. Some early Christian theologians attempted to save the everlastingness of God by assuming that Jesus was merely adopted as God's Son and was not truly divine. Accordingly, when Jesus was crucified, he was existing in time, but his timeless divinity allegedly remained undisturbed. God, in this kind of theology, did not enter into time after all. Another faction of early Christians agreed that Jesus is divine but that he only seems to have come into the realm of time, perishing, and death. Here again, God does not really suffer but instead remains timelessly unaffected by change of any sort.

A venerated early Christian writer, Irenaeus of Lyons (130–202), objected to both of these early attempts to keep God separate from time. He wrote: "By no other means could we have attained to incorruptibility and immortality, unless we had been united to incorruptibility and immortality. But how could we be joined to incorruptibility and immortality, unless, first, incorruptibility and immortality had become that which we also are?"[13] Likewise, the Nicene Creed, the fourth-century official codification of Christian beliefs, decidedly rejects attempts to separate God from our human experience of diminishment, perishing, and death in time. Jesus, the Creed insists, really lived in time and died an ignominious death in time. Yet, throughout, he remained "one in being" with God the Father.[14]

The unfading horizon of the not-yet, it seems, is what keeps cosmic time from being an illusion. And it is faith's openness to the not-yet that encourages us humans to live fully within time. Indeed, from a theological point of view, it is the coming of the future—the always newly arriving not-yet—rather than an imagined push from the past or a "fall" from eternity, that accounts ultimately for the reality of time and the reasonableness of hope. I will develop this (quite biblical) point more fully as we move into subsequent topics.[15]

It was Einstein, I want to emphasize here, who demonstrated that the cosmos can never again be separated from time. Einstein, however, failed to notice the dramatic implications of linking time and matter together. Time remained for him a dimension of his geometric understanding of nature rather than an irreversible narrative flow from past to future. Past, present, and future were for Einstein geometric facets of a single space-time universe. As we shall see later, it was his deeply religious love of eternity that persuaded him to absorb time—present, past, and future—into a standing present (*nunc stans*). Other scientists helped refine his geometry, but Einstein balked at the idea that cosmic time is irreversible and therefore that the universe can be understood as a long and still-unfinished story. He eventually changed his mind, but only reluctantly.

So we shall have to go beyond Einstein's personal understanding of time and eternity. We shall look for divine transcendence not so much in an eternal present as in an inexhaustible future that is still coming and hence not yet fully present. We shall look for God not just "back there" or "up there" but mostly "up ahead"—on the horizon of the not-yet. Indestructible rightness, in that case, takes the flux of time and whatever happens in time into itself endlessly and irreversibly. God is not outside of time, but instead God is the Absolute Future to which the universe is now awakening.[16]

Einstein was not prepared to accept the irreversible passage of time as objectively real. He thought our sense of duration was a subjective illusion having little to do with the nature of the universe itself. Not all cosmologists agree. An emerging cosmological opinion now gives new realism to our sense of the passage of time. The physicist Lee Smolin, for example, says that he has come to believe that "the deepest secret of the universe is that its essence rests in how it unfolds moment by moment in time."[17] Once we acknowledge, in contrast to Einstein, that time is an irreversible passage of moments, the synthesis of matter and time allows us to realize at last that the universe is a story rather than a solid, static, and purely geometric state. This adjustment provides the opportunity for a theological revision of our sense of divine transcendence as well. After Einstein, the cosmos is not merely spatial but spatiotemporal. If the world were merely spatial, then God's transcendence would mean that in some sense God is "not-here." But since the world is a temporal passage, to say that God transcends the world must mean that in some sense God is not-yet. And if God is not-yet, there is still room for hope.

Summary

Matter, Einstein discovered, is inseparable from time. For Einstein, however, our sense of the *passage* of time is an illusion. But what if time is a real, irreversible series of moments? Then time is a courier of stories; and the essence of stories, as we shall see, is to carry meaning. After Einstein, the universe looks like a story, but does that story have a meaning? Does it offer reasons for hope? In this book I attempt to answer these questions by engaging with Einstein on questions about God, eternity, time, and other important topics.

Christians' belief in God is part of the perennial human longing to link our fragile existence to a rightness immune to perishing. Faith in God reflects a common human conviction that only a connection to what is indestructible can relieve our anxiety about death and give permanent meaning to our lives. This is why the Abrahamic traditions have passionately protected the notion of divine everlastingness. Yet, as Christianity emphasizes, indestructible rightness has come into time irreversibly. If God remained altogether outside of time, then the world—with its perpetual perishing, suffering, and anxiety—would remain unhealed and without hope. But if God comes into time, and time flows into God, everything changes.

TWO

Eternity

> I believe in Spinoza's God, Who reveals Himself in the
> lawful harmony of the world, not in a God who concerns
> himself with the fate and the doings of mankind.
> —Albert Einstein

> My comprehension of God comes from the deeply felt conviction of
> a superior intelligence that reveals itself in the knowable world. In
> common terms, one can describe it as "pantheistic" (Spinoza).
> —Albert Einstein

> . . . and his kingdom will have no end.
> —Nicene Creed

ALBERT EINSTEIN LOVED ETERNITY more than time. Throughout his life he failed to see time going anywhere or leading to anything truly new. Time was a dimension of his spacetime universe, but for Einstein time was not a stream of moments moving from a fixed past to a new future. Perhaps his greatest contribution to science was to show that time is part of the fabric of nature, inseparable from matter itself. But time, he thought, is simply a geometric dimension of the cosmos, not an irreversible passage. The universe, to his mind, was not a directional current moving from past to future but a stationary "block" with no real distinction between past, present, and future. In his famous disputes with the French philosopher Henri Bergson, Einstein even insisted that our typical sense of time as an irreversible process is a "subjective" illusion. Time exists, he agreed, but it does not flow.[1]

Most of us find the idea of unflowing time hard to accept. A fundamental human experience is the one-way passage of time, the fixity of the past, and the prospect of eventually perishing in the future. For Einstein, however, time was not going anywhere. It may *seem* that we are being transported by time irreversibly from past to future, he allowed, but this impression is an imaginative fiction. Even today many physicists agree with Einstein. They endorse the notion of a cosmos in which time's passage is collapsed by mathematics into a timeless present indistinguishable from past or future. We live, they say, in a four-dimensional spacetime universe, but time (the fourth dimension) is not adding up to anything truly new. This denial that time is a real transition from past to future has made it hard, if not impossible, for Einstein and other scientists to understand the cosmos overall as a story of gradual awakening.

Fortunately, not all scientists agree with Einstein on this point. They are happy to embrace Einstein's discovery of the inseparability of time and matter, but they consider our commonsense experience of time's irreversibility to be an essential aspect of cosmic reality, by no means an illusion. In this book I follow the latter interpretation, but not for theological reasons. Theology, after all, cannot be a source of scientific information, nor can it adjudicate scientific disputes. I accept the objective reality of irreversible time because not to do so leads, as we shall see, to intellectual incoherence. The denial of irreversible time, I believe, is one of many ways to escape the anxiety of perishing. So Einstein's block universe is a new version of an ancient time-denying approach to nature that has snuck down the ages, misshaping philosophy, science, and Christian theology along the way. How the suppression of time's passage has affected science, philosophy, and theology is a central topic in these pages.

Even after the publication of his works on special and general relativity (1905–1916), Einstein's thoughts about time stayed much the same as before. And yet, by building time into his concept of nature, he unwittingly opened up a new future for the universe. It is Einstein's theory of gravity, as set forth in his general theory of relativity, that provides the mathematical infrastructure of Big Bang cosmology and of our present sense of the cosmos as a still-unfinished story. And it is Einstein's calculations, carefully interpreted, that, contrary to his own preferences, have set the universe free to be a drama, one whose secret meaning now lies hidden, not in eternity, but in the not-yet of irreversible time.

I refer to the post-Einsteinian universe as unfinished. I am not suggesting that the cosmos is headed toward a predetermined ending but only that the cosmic process is still under way and the future still open. In his youth Einstein had taken the world to be eternal and unchanging. In tacit obedience to his favorite philosopher, Baruch Spinoza (1632–1677), he assumed that nature has been around forever and that the laws governing it must be equally eternal. Like Spinoza, Einstein was convinced that the world's future, no less than its present and past, has been fixed forever. Although he never claimed to have a scholarly grasp of Spinoza's thought, he shared the philosopher's belief that the universe is both unbegotten and eternal. By wrapping the cosmos in the classic theological apparel of timelessness and necessity, he took it for granted that nature is all there is, that matter has existed forever, that the laws governing the cosmos are irrevocable, and that the universe simply *has* to exist.

Einstein was never able to accept the idea of a God who creates the world freely out of sheer goodness and who allows for its transformation in time. I believe that Einstein's love of eternity explains why he was disturbed at the prospect of a universe that undergoes major changes. Even after other mathematicians had corrected his equations, and after working astronomers had produced evidence of an expanding universe, Einstein did not abandon his affection for eternity. While his scientific work was reconnecting the cosmos to time, he remained personally enchanted with timelessness. He wanted nature to be eternal, and this longing, in part at least, is why he could make no place for a beneficent personal God who can make the world new.

Throughout his lifetime, whenever he spoke of God, as he did on more than one occasion, Einstein was not thinking of a responsive principle of love, fidelity, promise, and redemption distinct from the world, as the Abrahamic traditions have taught. He was expressing his reverence for the eternal mystery of a universe whose outer face conceals an unseen intelligence beneath it all. He relished and reverenced the mysterious fact that the universe is comprehensible, and that was enough, he thought, to qualify him as a religious person.

In writing or speaking about God, Einstein had in mind not an interested "Father almighty" but the mysterious rational orderliness of nature that makes scientific inquiry possible. Nor did he hesitate to give the name "faith" to his trust in nature's remarkable intelligibility.[2] If you look into the secrets of nature, he said, you will find beneath its surface "something subtle, intangible, and inexplicable." True faith consists of "veneration of this force beyond anything that we can comprehend." And so, Einstein concluded, "to that extent I am, in point of fact religious."[3]

Einstein's faith was a "cosmic religious feeling" for the timeless mystery of intelligibility underlying all appearances.[4] This feeling was essential to launching and sustaining the whole scientific enterprise. Faith was not "belief without evidence," as many scientific skeptics understand it, but a firm trust in the world's underlying comprehensibility.[5] Science, Einstein insisted, cannot get off the ground without faith of this sort. Carrying out a scientific research project requires a basic human confidence that the visible universe is the outward manifestation of a mostly invisible comprehensibility. We cannot say why the universe is comprehensible, so we need to take it on faith. Our minds do not create the universe's intelligibility but are instead obedient recipients of it. We can only marvel that the cosmos was already intelligible long before human minds came along to greet it.

Einstein saw something eternal, almost divine, about the comprehensibility that lies beneath appearances, so he would hardly have called himself godless. He believed in what he called Spinoza's God. Spinoza, however, had identified God, not with the liberating and responsive God of the Bible that his Jewish ancestors had worshipped, but with the universe itself. "God" and "nature," he had taught, are two words for the same totality. Identifying nature with God is known as pantheism, and Spinoza is its chief modern representative. Composed of two Greek roots, "pantheism" means that all (*pan*) is God (*theos*). So God is nature and nature is God. "My views" Einstein admitted, "are near those of Spinoza."[6] Einstein may not have been a strict pantheist, but like Spinoza, he denied the existence of any deity who exists independently of nature. Nature is all there is, but it has a face that makes it look divine.

The Appeal of Pantheism

Early Christianity, as the Nicene Creed illustrates, uncompromisingly rejected pantheism and by implication all other forms of time-denying naturalism. Christian faith insists that God is not nature but instead nature's creative ground and final destiny. Both Jewish and Christian teachings have officially rejected Spinoza's beliefs as inherently atheistic, but pantheism is religiously attractive to its adherents. It is appealing because it is healing. In one swift movement of the mind, pantheism absorbs nature, time, and human history into eternity. It links our fractured world so tightly to timelessness that wrongness is dissolved at once in the balm of nature's indestructible rightness.

Pantheism, like countless other forms of religion, expresses a passionate human longing to tie our perishable lives and the rest of the temporal world to an imperishable perfection. No less than other devoutly religious people, Spinoza had struggled to find courage in the face of fate, suffering, meaninglessness, and death. The most efficient way to satisfy this longing, he discovered, was to steep the cosmos so thoroughly in the sea of eternity that time was virtually dissolved. Einstein also thirsted for something like a pantheistic form of religious consolation. To understand his special attraction to Spinoza—and why he was not eager to accept irreversible time—it is helpful to remind ourselves just how restorative pantheistic piety can be to those who are unusually sensitive to time's corrosiveness. When a close friend of his died, Einstein wrote: "Now he has departed a little ahead of me from this quaint world. This means nothing. For us faithful physicists, the separation between past, present, and future has only the meaning of an illusion, though a persistent one."[7]

Pantheism, however, pays a price for eternalizing the cosmos. First, it snatches time up into eternity without giving the universe the opportunity to *become* something. Devoid of a sense of the not-yet, pantheism robs the cosmos of any real future and, in doing so, gives no cumulative significance to the passage of time. Thus, it casts doubt upon the unique, unrepeatable value of every moment in natural and human history. Second, by imprisoning our still-emerging universe in the stiff armor of eternal necessity, pantheism cannot allow the cosmos to be an adventurous drama of awakening. Third, pantheism's ideal of indestructible rightness is not self-sacrificing love but an obdurate protectionism that keeps the world from ever becoming new. And fourth, pantheism is willing to sacrifice human freedom on the altar of nature's predictability.

To Einstein the price seemed right. Although he was not a card-carrying pantheist, he treasured the idea that in principle whatever happens in nature is fully predictable, including human activity. Einstein's famous dictum that "God does not play dice with the universe" implied for him that the laws of nature are inviolable, that the cosmos is virtually finished, and that humans, as part of the natural world, are not really free.[8]

Impatience

If we look at pantheism today in the light of an irreversibly temporal universe, it gives the impression of being an extreme instance of religious impatience. While biblical religion makes a virtue of waiting for God, the true pantheist has no reason to wait for anything. Everything real has already happened—eternally. Pantheism is seductive and soothing because it joins our lives and minds to eternity without having them sail through the narrows of irreversible time. By immersing the cosmos immediately in the solvent of eternity, pantheism makes everything that happens in nature and history—and in our own lives—inevitable.

Such a worldview can be very appealing if time seems to be an enemy rather than a supportive friend. The belief that everything that happens in the universe happens by necessity may also be attractive to scientists, since science relies on predictability. The almost irresistible lure of predictability shows up today in the guise of scientific determinism. This is the belief that inviolable natural laws have, from the start, fastened down everything that is going to happen in the universe. Determinism, though now suspect even to many physicists, is enticing because of its anxiety-reducing promise to expel all uncertainty from the sphere of true being.

Scientific determinists usually adopt the worldview known as materialism, the belief that lifeless matter is all that really exists. Materialists deny not only that freedom is real but also that a personal, responsive deity exists. The two denials go together. Both Spinoza and Einstein, nonetheless, would have found themselves ill-suited to the company of contemporary atheistic materialists. Each of these great thinkers had a profoundly religious feeling for eternity. Devotion to the theme of indestructibility permeates their thoughts from top to bottom. Unfortunately, their ideas of God—or nature—have the quality of rigorous causal necessity characteristic of modernity's well-crafted clocks and other machines. God for Spinoza was free only in the sense that divinity is its own cause, and to the pantheist this means that nature, since nature is God, is its own cause. But apart from the primary act of self-creation, whatever happens in nature, happens by necessity, according to Spinoza. In principle, everything is settled forever. Einstein, similarly, was convinced that determinism was inseparable from his four-dimensional spacetime universe. No room existed for true novelty, chance, or freedom—or for irreversible time. So, he thought, there could be no inherently unpredictable future. Timelessness alone was real, which ruled out the possibility that nature was still open to what is truly new.

Einstein's fixation on eternity, like Spinoza's, was profoundly religious. Yet his own science of relativity, as it has been examined and interpreted by cosmologists ever since 1915, casts serious doubt on the idea of an eternal and necessary universe. Relativity provides the theoretical foundations of Big Bang cosmology, according to which the universe has existed for a finite amount of time rather than forever.[9] If relativity is right, and if time is both real and finite, the universe cannot be eternal and necessary. It is temporal through and through. It has a contingent beginning and an uncertain destiny. Its itinerary has already proven to be one of unpredictable twists and turns. Nature is not a machine but an awakening.

Science now shows that the universe is still coming into being. The most precious emergent outcomes of cosmic history so far—life and mind—could not have been predicted by a deterministic analysis of the early universe. Rather, they are developments in a cosmic awakening with no fixed end in sight. Awakening to what? This is a way of asking the main question for theology after Einstein. To find out what our unfinished universe is all about we have to tune into what is not-yet. We cannot be too impatient when trying to determine where the cosmos is going or whether it has a fixed destiny. We cannot expect an unfinished universe to be fully intelligible here and now. We need to be patient and wait.

Spinoza provides the most sophisticated philosophical argument modern thought has yet made for the futility of waiting. He differs most from his biblical ancestors by identifying the physical universe with eternity and necessity, traits that theology traditionally associated with a timeless God. In effect, Spinoza denies that there is real temporal movement in nature. His pantheism is heir, at least indirectly, to the ancient Platonic insistence on the timelessness of God. Accordingly, if with Spinoza we identify nature with God, nature must be eternal and necessary. Not only would God thereby be immune to any real contact with time but so would the universe.

But is the universe eternal and necessary? Spinoza's way of thinking, as I have suggested, satisfied Einstein's religious sensibilities, which helps to explain why the great physicist was not prepared to accept the universe as an irreversible temporal passage. Even after friendly critics pointed out to him that his spacetime cosmos need not have been the same forever, Einstein clung as long as he could to his belief that the physical universe was eternal and necessary despite locally changing appearances. He even tinkered with his equations to make them fit his personal—one might say religious—preference for a static cosmos. In the face of expert objections, Einstein went looking for a place in his complex set of numbers for a "cosmological constant," a mathematical fudge-factor that would keep the universe from changing in any significant way. After being shown the mounting new astronomical evidence of an expanding universe, he had to change his mind, of course, but he seems never to have fully abandoned his Spinozist predilection for timelessness.[10]

When, in 1921, the Archbishop of Canterbury asked Einstein about the religious significance of his theory of relativity, the physicist replied curtly that there was none, unaware that his preference for a static universe was itself already theologically loaded.[11] The archbishop's question was not silly. The theological implications of Einstein's scientific cosmology are profound. Revolutions in science always affect our ideas about God whether or not we believe God exists. During the early modern period, for example, when philosophers of nature started picturing nature as a machine, theologians assumed, almost unconsciously at first, that God must be something like a mechanic, watchmaker, or "intelligent designer," a dubious association that contributed much to the rise of modern atheism.[12]

After Charles Darwin had cast doubt on the belief that living organisms are the product of direct divine engineering, the idea of a designing deity became increasingly unbelievable. A mere half-century before Einstein became famous, Darwin had already demonstrated that the specific features and endless varieties of life on Earth are the product not of eternal necessity but of an unpredictable historical process. The recipe for biological evolution, Darwin theorized, consists of three ingredients: first, *accidental variations* in natural history and in the inheritance of biological forms; second, the impersonal filter of *natural selection*, the "law" that discards all organisms that cannot adapt and survive long enough to reproduce; and, third, an enormous *passage of time* during which the combination of accidental changes and blind natural selection is given sufficient opportunity to bring about all the diversity of life on Earth in a gradual way. Life, then, is not the predictable product of timeless necessity but the outcome of an interweaving of contingency, regularity, and irreversible time. It so happens that contingency, lawful regularity, and the passage of time are also three essential elements in the makeup of any story. The story of life, therefore, because it has all three ingredients, is an ongoing production for whose specific outcomes we shall always have to wait.

Darwin's own Christian faith slipped away from him as time and contingency supplanted eternity and necessity in his thoughts about nature and God. Indeed, as a result of Darwin's new story of life, many educated people to this day seldom think about God at all. After Einstein, however, we may have new thoughts about God and eternity because we are compelled to think new thoughts about time and the natural world. We now know that nature is not eternally fixed and frozen. Never again may we plausibly fit the adjective "timeless" to the cosmos as pantheism does and as Einstein earnestly longed to do. Nature is a temporal process that has continued to bring forth surprising new results as time travels on. The universe is a display of neither engineering elegance nor a totally blind and aimless drift. It is an unfinished drama, by which I mean a suspenseful narrative that may carry a meaning for which we shall have to wait. We approach dramas, after all, not in search of mathematical intelligibility but in search of narrative coherence. Things are going on in the universe whose intelligibility can become apparent, but only if we give them time to unfold and then tell stories about them.

Both the evolution of life and the brief history of humanity on Earth now show up as unprecedented new chapters in a cosmic drama of awakening. The drama is much older, vaster, and more interesting than we had ever guessed before the twentieth century. Consequently, instead of obsessing about whether life manifests "intelligent design," it is now possible—and I believe theologically more fruitful—to ask whether anything of lasting importance is going on in the long cosmic drama. In *God after Darwin* (1999) and subsequent writings on theology and evolution, I have expressed my sympathy with Darwin's doubts about a designer-deity.[13] The most important theological implication of Darwin's discoveries, I have argued, is that life has a dramatic quality that, for all we know, carries a momentous meaning deep down, far beneath the surface samplings taken by the biological sciences. Likewise, beneath the soundings of astrophysics lies an adventure of awakening that stretches throughout the pilgrimage of matter in irreversible time. Einstein's greatest contribution to theology, in that case, is to have drafted the outlines of a universe whose intelligible format is not just geometric but, as we shall see, also dramatic.

Stars and Story

That there is a real connection between the heavens above and human life here below was an almost universal assumption of prescientific thought. Most people assumed that both the good and the evil things that happened to them were tied to the stars in a hidden way. The term "dis-aster" (literally, bad star, or ill-starred) reflects the ancient assumption that personal fortune or misfortune cannot be divorced from the alignment or misalignment of astronomical bodies. In the European world, looking to the stars and planets to interpret human lives was a passion that nearly everybody, including popes, indulged at times.

Today, however, it is not angels or "ether" that links us to the cosmos. It is matter's inherent narrativity. Relativity's conjoining of matter and time now allows us to tell a long story about how present phenomena, including our own lives and minds, are tied to the physics of the early universe. Scientists can now give an increasingly detailed account of how everything in cosmic time came to be connected narratively with everything else. The fact that nature is tied to increasing entropy—that is, to the increasing loss of energy over time—and hence to irreversible time, also allows for the enfolding of each person's biography into the supportive fabric of a long cosmic narrative that is still incomplete. Nature, or what theology calls creation, is an unfinished drama rather than the product of an opening instant of divine magic. This means that the universe is undergoing a suspenseful and anticipatory process into which the evolution of all species and the brief stories of our own lives are woven seamlessly.

It is because of the universe's dramatic makeup, rather than because the Big Bang may seem to support the idea of divine creation of the world "in the beginning," that Einstein's universe is theologically so interesting. The cosmos now shows evidence of having been at least loosely shaped from the start by what we may call a "narrative cosmological principle."[14] Built into matter, in addition to mathematical pattern, is the kind of indeterminate meaning that humans look for when reading or listening to stories.

Consequently, if the cosmos exists as an unfinished story going on in irreversible time, we have good reason to wonder whether its narrative constitution plays host to a kind of meaning that science cannot reach, a meaning that may respond to our human attempts to live with purpose and hope. What makes the post-Einsteinian picture of the universe so theologically compelling is that the kind of intelligibility it carries adds to the idea of geometric patterns a possible narrative disposition that may speak to our personal longing for a richer kind of meaning than that given by geometry.

Einstein, of course, was not looking for narrative coherence in the cosmos. He was following the habitual modern scientific ideal of seeking timeless mathematical intelligibility. This is a worthy intellectual goal—

and one that has helped shape scientific method since Galileo, Kepler, and Newton. But Einstein, like some of his predecessors in modern science, believed that the universe could be understood in depth only by geometry. He took for granted throughout his lifetime that the cosmos is governed impersonally by remorseless physical regulations and timeless mathematical principles. His habitual preference for a universe that exists by necessity, along with his denial that the passage of time is physically real, kept him from asking what else may be going on in the universe. Above all, he failed to allow that beneath its geometrical elegance lay something breathtakingly significant—a dramatic awakening to which his own life and work have themselves made significant contributions. After Einstein, the universe has shown itself to have a refreshing openness to the not-yet. This makes it a fit habitat both for endlessly questioning minds and for souls that cannot breathe in the absence of an open future.

The Flight from Time

Pantheism leaves no place for valuing the journey of time. The denial of time's narrative importance has also been a temptation for Christians almost from the beginning of their religious history. An impatience to escape from time showed up in early Christian heresies such as Gnosticism, Docetism, and Monophysitism. Though differing in many respects from one another, these deviant teachings shared what they considered a noble longing to protect indestructible rightness from being dissolved in the perishing of time. They did so by concocting alternative versions of what the New Testament means when it refers to Jesus as the Son of God.

In what sense, early Christians wondered, can Jesus be properly referred to as divine, and hence indestructible, if he is also a perishable human being? Christians had a hard time holding time and indestructibility together, as is reflected in the history of doctrinal disputes about who Jesus really is. Only with difficulty did early church councils curb the religious temptation to dissociate Jesus from time. They were not always successful. Like other religious people, including pantheists, Christians have sometimes loved eternity so much that they have sought to purify it of any contamination by time. Attempts to separate God from time are persistent throughout Christian history. They are still going on.

The Nicene Creed claims that God and time are inseparable. Though "begotten" "before all ages," the Son of God was born of a woman who existed in time. Official Christian faith has always emphatically denied that Jesus is a ghost. Moreover, he "was crucified" at a specific point in time—namely, under Pontius Pilate, as the Apostles' Creed notes. It was during an identifiable period of history that "he suffered death, and was buried." In Jesus, God was fully "incarnate" (or "enfleshed") in a human body and all that entails. The body of the historical Jesus, after all, was made of carbon and other elements forged in stellar ovens billions of years ago in cosmic time. That body was a product of the same physical and evolutionary processes as all other living beings.

Christians officially confess that attempts to wrest Jesus free of time and matter are heretical, but these efforts are nonetheless recurrent. The tendency to deny the reality of time is understandable, rooted as it is in our common human anxiety about losing touch with indestructible rightness during our brief span of existence. It is not surprising that so much religion, including some strains of Christian theology, links up with the religious instincts of Spinoza and Einstein. Sensitive to the fact that time entails perishing, Christians have felt an almost unconquerable temptation to make eternity real at the expense of time. In the interest of cleansing eternity of time, early Christians sometimes even tried to do away with the humanity of Jesus, dissolving it in his timeless divinity.

A notable example of this flight from time came with Docetism. Docetism (from the Greek verb "to seem") maintains heretically that the bodily existence of Jesus in time was only apparent and that the Son of God only seems to have suffered and died. Docetism, though existing in several versions, has been a constant temptation in Christian spirituality because, like pantheism, it appears to deliver the indestructible God from the dangerous linkage to the passage of real time. By exempting the Son of God from the perils of historical existence, Docetism allows the followers of Jesus to avoid living fully in the physical universe—a temptation that continues to diminish the vitality of Christian existence today. By uniting human souls with a dematerialized Jesus, Docetism comforts its adherents with the thought that they, too, are essentially unreachable by time and death. And by denying the doctrine of God's incarnation, Docetists implicitly deny that time itself has a meaning.

Nevertheless, although Einstein's religious instincts led him to side with the heretics, his mathematical physics has been updated so that it now entails a universe that exists irreversibly in time. What is most theologically significant about Einstein, then, is not his personal enchantment with timelessness, a penchant that may easily lead his admirers to extol his alleged mysticism. Rather, it is what his theory of gravity has to say about the importance of time. It is not Einstein's personal philosophical opinions but his revolutionary contributions to scientific cosmology that comport most comfortably with official Christianity's affirmation of time.

The Nicene Creed rejects any religious consolation that steers us away from the reality of time, since its inner message is that God does not avoid time either. The Son of God, according to the official teachings of Christianity, truly entered into time—to the point of suffering and dying. This means for the rest of us, too, that contrary to the countless religious denials of time, we cannot experience indestructible rightness apart from our own living, enjoying, striving, suffering, and dying in time. To believe in God means to experience time fully, in solidarity with the crucified Christ and his irrevocable connection to the history of matter.

In developing its strange revolution in the idea of God, early Christianity began to formulate a doctrine of the Trinity on the basis of allusions in the New Testament to God as Father, Son, and Holy Spirit. No one has ever fully clarified the meaning of these mysterious references, but a fundamental objective in the shaping of Trinitarian theology has been to protect the idea of a God who loves time from being lost in it, while also emphasizing that indestructible rightness does not separate itself from time by hiding in eternity. The history of the doctrine of the Trinity reveals a religious community's struggles during two millennia to hold on to the paradox that the rightness manifested in Jesus's life and death remains indestructible, not by happening outside of time, but by dwelling in, embracing, and renewing it.

Consequently, indestructible rightness cannot for Christians mean immunity to time, suffering, and death. The Nicene Creed and the doctrine of the Trinity imply that time does not lie outside but inside the life of God. Yet the longing to separate God from time still hovers over Christian life and theology. Early on, out of a concern to preserve the imperishability of God, Christian theology formed an alliance with Platonic thought that unfortunately it has never gotten over. This union gave rise to a kind of theology that for centuries has led people of faith to avoid squarely facing the paradox that time is inseparable from indestructible rightness. In view of cosmological developments after Einstein, however, it seems intellectually incongruous for Christians to separate their religious quest for salvation from the fundamentally temporal orientation of the whole universe. The main theological issue now is not whether patterns in nature point to eternal forms in the mind of God but whether the story of the universe is carrying with it a momentous promise yet to be realized.

Summary

Einstein's theory of gravity, as set forth in his general theory of relativity, provided the mathematical underpinnings of Big Bang cosmology and our present sense of the cosmos as a still-unfinished story. Einstein, like the pantheist philosopher Spinoza, loved eternity more than time. Yet it was his own science that, contrary to his preferences, helped set the universe free to be a temporal drama, the meaning of which lies hidden in the not-yet. The Abrahamic God is one "who makes all things new." So with each passing moment the power of the not-yet delivers the universe from fixity to the past and gives it a new future. Christian theology's interest in Einstein, therefore, lies not in his personal, almost mystical love of eternity but in his laying out—with the help of other scientists and mathematicians—the outlines of a universe that exists, awakens, and evolves in real time.

Christians, as instructed by their Creed, believe that the luminous figure killed at Calvary is now alive, cosmically as well as spiritually. The incarnate Word, therefore, does not lead souls away from time but leads all of time toward a fulfillment in God. In my opinion, the Nicene Creed, of all ancient codifications of philosophy and religious faith, is unmatched in its affirmation of the reality, irreversibility, and significance of time. The Creed is not a map for escapist flights into eternity but an encouragement to embrace the reality of time fully. In principle, at least, Christian faith takes time more seriously than most scientists and philosophers do. The Creed thereby justifies our love of nature because it affirms that the "maker" of "all things, visible and invisible," has fully embraced time, guaranteed its irreversibility, and promised to save it everlastingly.

THREE

Time

Henceforth space by itself, and time by itself, are doomed to fade away into mere shadows, and only a kind of union of the two will preserve an independent reality.
—Herman Minkowski (Einstein's professor)

It is nonsense to conceive of nature as a static fact. There is no nature apart from transition, and there is no transition apart from temporal duration.
—Alfred North Whitehead

All of the mysteries physicists and cosmologists face—from the Big Bang to the future of the universe, from the puzzles of quantum physics to the unification of the forces and particles—come down to the nature of time.
—Lee Smolin

PRIOR TO THE TWENTIETH CENTURY, neither scientists nor theologians had any idea of time's immensity. Most people familiar with the creation stories in Genesis assumed that humans were present in nature almost from the beginning and that the beginning was not so long ago. Strict biblical literalists still think the universe is only several thousand years old, and even scientifically educated believers seldom reflect earnestly on the religious meaning of the enormous spans of time through which the universe has journeyed so far.

Maybe we need more time to think about it? It was only a little over two centuries ago, after all, that geologists began gathering unambiguous evidence that planet Earth has a long history, vastly greater than their biblically informed ancestors had imagined. Only during the past century and a half have biologists come to realize that life on Earth is many millions of years older than even Darwin had suspected. And only during the past century have cosmologists found out that our Big Bang universe is billions of years old, with—quite likely—many more billions to go.

In religious thought prior to the age of science, time sometimes seemed like a skin that the world needed to shed, not a sequence of opportunities for the building up of value and meaning in a drama of awakening. As far as human experience is concerned, time seemed to be a stain from which souls needed to be cleansed. The relentless passage of time, it was thought, was the cause of perishing, destruction, decay, and death, so why not flee from it or abolish it altogether?

Even today the ideal state of existence, according to most religious seers, is that of timelessness. Experts in spirituality tell us we can escape the ravages of time and the terrors of history by returning—by way of contemplation, sacred rituals, and eventually death—to an original plenitude residing outside of time's natural flow. We can experience inner freedom only by allowing our souls to be swept out of time and into eternity. We must be saved *from* time, since we do not fully belong to the temporal, physical universe. Matter and time, in this understanding, must be shuffled off while the soul is making its "journey into God."

Not to be outdone, many scientists do not wish to get too cozy with time either. A fair number of them assume with Einstein that the sense of time passing is a psychological illusion.[1] Mathematical physicists sometimes consider time virtually nonexistent in comparison with the timeless Platonic bath of pure numbers into which the temporal universe dips for its intelligible forms.[2] In the preceding chapter we observed that Albert Einstein was not in love with time either, and almost seven decades after his death scientists are still looking for ways to get rid of it. For example, the typically materialist reading of nature effectively does away with time by reducing the whole story of the universe to the virtually timeless subatomic plasma from which it all began. One way of escaping time, they assume, is by coming to rest at its beginning.

In 1915, Einstein had not yet realized that his impressive theory of gravity (published in 1916) could be interpreted to mean that the whole universe follows an irreversible arrow of time from past to future. The clearest evidence of time's irreversibility, however, is not to be found in relativity but in thermodynamics. The heat-energy lost in doing work—known as entropy—is irrecoverable, which means that there is no going back in time. As the physicist Carlo Rovelli points out, "This is the only basic law of physics that distinguishes the past from the future." Remarkably, it is the loss of energy available to do work—the increase of entropy—that "powers the great story of the cosmos."[3]

Einstein demonstrated that time and matter are inseparable, but for him and many other interpreters of relativity, time does not really flow from past to future. Time, they assume, is a fourth dimension of a spacetime universe, not an irreversible series of moments. Throughout his career Einstein privileged timelessness over time. Yet his mathematical science, which needs to be distinguished carefully from his religious ideas and philosophical leanings, can be interpreted to mean that the whole of nature is an irreversible drama of temporal transformation.

Today the standard Big Bang theory—based on Einstein's theory of gravity as interpreted by Georges Lemaître and others—implies that originally, at the very beginning of time, the universe was an infinitesimal physical speck smaller than an atom. During the first instant of the universe's existence the immeasurably hot and dense primordial smidgen began to cool and expand, which it has continued to do for 13.8 billion years. During its long passage the universe has gradually brought forth atoms, stars, galaxies, and—at least on Earth—living cells and diverse organic species. This narrative of increasing complexity is another sign of time's one-way movement from past to future. During the past several million years the nervous tissue and burgeoning brains of several terrestrial lines of life have become increasingly conscious. In the human species consciousness has intensified to the point of aspiring to indestructible rightness.

This whole story of awakening rides physically on the universe's irrecoverable loss of usable energy. A thermodynamic price is always being paid as the story moves along. The long cosmic journey has gone on without violating any so-called laws of nature, and scientists rightly assume that nothing can happen in cosmic history that breaks or bends the regular routines of physics and chemistry. Nature's fidelity to strict rules, however, does not mean, as we shall see, that the cosmic future is already set. While nature follows the same inviolable physical regulations from age to age, this consistency leaves plenty of room for a cosmic drama, an adventure that is new every moment.

Einstein did not see it this way. He noticed no dramatic story, only the laws of nature working the same way from age to age. Even more emphatically than most of his fellow scientists, he insisted that nature's rules are fully determinative of everything that happens. He thought so, however, not only for scientific but also for religious reasons. He considered the laws of nature to be timeless, and out of a devout love for timelessness, he revered their immutable ordering presence almost as though it was sacred. Although most interpretations of his general theory of relativity allow that the physical universe changes considerably over long periods of time, this supposition did not offset Einstein's personal conviction that the cosmos is ruled by unbending necessity. The same great scientist whose mathematics set the cosmos free to exist in time was personally content to lock it up inside a geometry that would keep it from ever having an open future.[4]

Today a significant number of cosmologists have come to believe that the cosmos, though sticking to timeless rules, flows irreversibly from past to future. Because the overall loss of heat energy in the universe is irrecoverable, the universe will end badly, in thermodynamic terms. In the words of the physicist Katie Mack, "The current scientific consensus is that far in the future, long after our own sun dies, whether or not we have evolved from fragile biological humanity to spacefaring, self-replicating machines, we will all eventually succumb. The distant galaxies will be pulled beyond the limits of our sight; stars will die; all light will fade. We will end our cosmic existence alone, in the dark."[5]

Einstein, with his fixation on eternity, did not seem to worry about such an outcome. His theory of relativity forbade the separation of time from matter, but he thought of time as static rather than corrosive. In his view, the timelessness of nature's unchanging laws could not be defeated by time. Einstein, of course, is by no means alone among scientists in failing to allow for the passage of time and a place for real novelty in the universe. Countless others also believe that everything that happens in nature is the inevitable result of physical causes obeying rules fixed forever. On this basis, scientific celebrities have rejected, along with time, the possibility of human freedom. Fortunately, as I argue later, there is no good reason for drawing this conclusion. Nature is a coming-to-birth; what it will be in the future nobody can say exactly, even in principle. Although its physical rules may not change, the story that weaves itself into nature's routines spins off countless surprising outcomes that cannot be perfectly pictured in advance.[6] Dramatically speaking, the cosmos is a series of contingent events in which new kinds of beings emerge in the pattern of unprecedented narrative arrangements—and without breaking any physical laws. Accordingly, nature's deepest intelligibility is not that of mathematical simplicity but of a dramatic coherence not yet fully actualized.

Although the universe may run out of usable energy, and everything in time will perish, the story of the universe need not be condemned to absolute oblivion. It will always be true, even after everything else in our universe is gone, that the story of the universe, and the story of your life and mine, took place at a specific time in natural history. Trillions of years from now, after the sun has died and the universe has run out of steam, it will still be true that you and I lived and that we were part of a grand story. But where will it still be true? Theology's answer: In the everlasting, preservative care of a God who gathers up, embraces, and redeems the full cascade of events in time. We shall examine this response in Chapter 11.

Although Einstein sticks to the classical deterministic philosophy of nature, his theory of general relativity can be interpreted to mean that the cosmos is temporal, not eternal. Its temporal flow allows it to carry a dramatic content hidden from astronomy, physics, and chemistry. The universe has already blossomed into scientifically unpredictable outcomes— like Einstein's mind. In fact, the existence of minds—yours, mine, and Einstein's—is all the evidence we need to realize that the universe, dramatically speaking, is a wondrous adventure of awakening. The activity of your own cognitive life, for example, as you are now probably questioning what I have just written, makes the point irrefutably. In view of discoveries in physics, biology, and other sciences, you cannot plausibly deny that your mind is part of the universe. And you cannot assume that your thoughts are happening somewhere outside of nature and time. Your cognitive life, viewed cosmically, *is* the universe now awakening in a unique and unrepeatable way.

The phenomenon of thought—as it is activated in your own insights and questions at this moment—is an indication that the universe is still awakening. However, no wealth of present or past scientific analysis could ever have predicted the shape of your questions, or the content of your thoughts, or the moral choices you will be making today. That you are now entertaining ideas and sentiments that have never occurred before in the history of the universe does not mean that your mind is leaping outside of time or violating any biological, chemical, or thermodynamic regulations. Rather, the present content of your mind and your unique awakening to meaning and truth demonstrate that the universe that gave birth to that mind is inherently open to an unpredictable future. Not everything is set in stone from the start. Contrary to Spinoza's and Einstein's philosophical preferences, discussed earlier, time is not an illusion, and the universe is not the same from age to age.

Again, it is the irreversible movement of time that allows the universe to be a drama rather than a fixed design. Before Einstein, time seemed to exist independently of local events. Isaac Newton, for example, assumed that every now in nature existed simultaneously with all others in universal time. If an event took place at a certain moment on Earth, imaginary observers could register it as occurring at the same time everywhere else in the universe. Einstein's science of relativity, with its denial that time is independent of matter, does away with the illusion of universal simultaneity. Time is not a container or a fixed tunnel through which the universe journeys. It is part of the physical makeup of nature itself. Take away time, and nature as we know it no longer exists. Take away irreversibility from time, however, and you end up with a pointless universe.

In 1905, Einstein published his theory of special relativity, demonstrating that everything in nature and time is conditioned by the speed of light. The speed of light remains constant while everything else, including time, must bow to the limits it sets. In 1916, Einstein went public with his general theory of relativity, demonstrating the equivalence of acceleration and gravity. He theorized that in the presence of massive bodies time slows down only to speed up as it escapes from them. His theory has now been confirmed—the accuracy of GPS technology, for example, is based on it. Time can no longer be divorced from nature.

Irreversible time is a necessary condition for the existence and telling of stories. We humans express the meaning of our lives and other series of events by telling stories about them. The search for a story's meaning is a search for a unifying principle or set of events that promises to tie together multiple events and episodes in time, giving them an overall intelligibility that they would not have in isolation. If time is not real, then the human search for nature's intelligibility may be satisfied by geometric rather than narrative coherence. For mathematical thinkers from Pythagoras to Einstein, geometry provides intelligibility enough. But if time is irreversible, then the searchers for intelligibility are obliged to look for a principle of meaning, or perhaps a story, that ties together the series of moments that take place in time. After Einstein, this is where theology may come into the picture. Theology looks for narrative meaning and hence for what I shall be calling dramatic coherence. In addition to the illuminating scientific search for geometric coherence, there is room now—after Darwin and Einstein—to look deeper than geometry for dramatic coherence in the cosmic story.

We humans, unlike other living species, have an exceptional craving for meaning, and it is the function of stories to tie our experiences together into a dramatic coherence. After Einstein the whole universe for the first time shows itself to be a long and still-unfinished drama. But does it also carry a meaning?[7] If so, the meaning would arrive in the format of stories that we humans seek and tell. May we not do so also with the universe itself? Is it not possible that the cosmic story is pregnant with dramatic meaning alongside, and embedded in, its geometric intelligibility?

In its quest for understanding, human consciousness approaches the world with something like a narrative a priori. Deep inside our genomes, nervous systems, and inquiring minds lies a story-shaped cavity within which the content of our experiences accumulates over the course of time. It is not that our minds create narrative patterns out of nothing and then project them onto a blank, storyless universe. Rather, the inherently narrative quality of the universe is imprinted in the deep structure of each mind to which the universe has given birth. It is natural, then, that our native quest for intelligibility would lead us to look for dramatic coherence in the universe. The universe's dramatic meaning, however, is always, in some way, not-yet.

Something about the very genre of story, furthermore, is soothing and healing. We can put up with anything, including a dangerous universe, if we can fit it into a story.[8] It is not surprising, then, that religions almost universally express their sense of meaning in the form of sweeping narratives, sometimes called myths, many of them cosmological. These fascinating stories tell us about the world's origin and destiny. In doing so, they locate our experience in a wider scheme of things than everyday awareness allows. Sacred myths guide our conduct, and give us a sense of reality. They also tell about the origin and end of evil and hence about rightness and wrongness. By passing on sacred stories from one generation to the next, religions have been able to tame the terrors of time and bring broad intelligibility to what would otherwise be frighteningly obscure.

Likewise, the new scientific cosmic story may appeal to our longing for comprehensive narrative meaning. Unlike religious stories, however, the cosmic story takes place in real time. All events that happen in time can be given at least a minimal kind of intelligibility by being linked to other events in an encompassing story. It is the nature of stories to reveal connections for the sake of deeper meaning. The universe story, too, as it turns out, unfolds in the pattern of a long narrative. Maybe the universe journeys on in accordance with a narrative cosmological principle. In addition to geometric coherence, it may be carrying a dramatic meaning, one that can provide reasons for hope.

The Real Story

Einstein's general theory of relativity provides the basis for the contemporary consensus among most scientists that our universe originated in a Big Bang billions of years ago. And even though the universe is not eternal, as Einstein's Spinozist religious instincts would have preferred, it has been around for a long time. To get an idea of its immense temporal scale, picture a bookshelf containing 30 large volumes. Each book is 450 pages long, and each page stands for 1,000,000 years. The cosmic story begins 13,800,000,000 years ago on page 1 of volume 1. During the first twenty-one volumes no living beings are present in the universe. The earliest sparks of life do not begin to glow until volume 22, and they do not billow into complex multicellular organisms until around the end of volume 29. Life is not in a hurry to come into the universe. During the Cambrian period, starting around 570,000,000 years ago, and lasting for several million years, life "suddenly" grows more complex than before. The fossil record deposited during this remarkable chapter in terrestrial history puts on display the remote ancestors of many present forms of life, including human beings.

Dinosaurs start arriving on Earth a little after the middle of volume 30, and they go extinct 66,000,000 years ago, around page 384. This is when mammals, already present in the age of dinosaurs, begin flourishing. Fifty-five pages or so from the end of volume 30 we come across the earliest primates. Humanlike mammals start showing up during the final few pages of the last book, but anatomically modern humans do not come into the story until the bottom of the very last page of volume 30. Only in a small paragraph at the end of volume 30 do we read about the arrival of organisms capable of reflective thought, ethical aspiration, and religious restlessness—at least on our middle-sized planet.

The long expanse of time prior to the origin of life and thought on Earth may seem like an enormous waste, especially if your theological instincts prefer the kind of creator who can design living organisms and give them complex brains on page 1 of the first volume. Why, you may wonder, did life and mind take so long to show up? Why, if the universe story has a point, is it taking so much time to reveal it? What theological meaning, if any, can be wrested from all this waiting around?

Scientific skeptics reply that it took so long for life to arrive because there is no creator God putting things into existence. Behind the emergence of life there is no intelligent divine designer, only blind chance and impersonal physical laws. If God existed, the doubters assume, then the universe, life, and mind would have sprung into fullness of being in a single act of miraculous inventiveness at the beginning. But, as science has demonstrated, that did not happen. Hence, secularist skeptics insist that what many religious people call God cannot possibly exist: the vast amount of time it took for matter to come alive, and for life to become conscious, is proof of it.[9]

The skeptics, however, are looking not for a creator but for a magician. So, too, are many religious believers, which is why they reject or ignore the evolutionary and cosmological discovery of deep time. The universe must either be frozen eternally, as Einstein and Spinoza postulated, or, if it had a beginning, it should have sprung fully into being in an impressive opening instant, as many Christians still prefer. A barely concealed hostility to time underlies both opinions.

New knowledge of time's deliberate pace tells us that the world's creation, attributed to God in the Nicene Creed, is not a feat of divine engineering proportioned to our impatience and our narrow human sense of decent design. Instead, creation is the story of a gradual, elongated evocation of unpredictable new forms of being. Even the chemical components of life are not fashioned in a fixed way on the first page of our thirty volumes. It takes many billions of years for the universe to arrive at a state in which carbon atoms can come into the story and, along with other chemical elements, provide the building blocks for the slowly emerging architecture of life.

There is nothing magical about any of this. Instead, there is something even more wondrous—namely, that irreversible time makes matter dramatic from the start, opening up the possibility that the cosmos is a story that may be carrying a dramatic meaning. Science cannot see any of this because scientific method is not wired to discern meaning, value, or importance. This is no disparagement of science, since its method of understanding is to look for intelligibility in the form of mathematical, statistical, or geometric arrangements. It is no put-down of science to observe that it cannot look deeply inside the cosmic story any more than it can look into the story of your own life. When you tell your own life-story, you are recounting the close calls, near misses, adventures, joys, and regrets associated with its various episodes, and you are doing so by way of personal knowledge, not through the detached method of scientific understanding. That is, you reveal what's going on in your life in the form of dramatic rather than geometric meaning.[10] It is the same with the cosmic drama. To get a sense of what's really going on, outside observation and mathematical measurement are not enough.

There is an inside cosmic story to tell, one that only a dramatic frame of mind can gather up and deliver to human attention. From the perspective of thermodynamics, the universe is a purely physical process going through a gradual, measurable, and irreversible loss of usable heat energy. From another perspective, however, the same universe is a momentous drama of awakening. The significance of time, in that case, is that it allows matter to take on the shape of a story whose intelligibility still lies mostly out of sight—in the hidden arena of what is not-yet.

From the perspective of a physicist, time is what keeps everything from happening at once.[11] From a dramatic perspective, time is what allows matter to be a story. Without contradicting Einstein's focus on the geometric structure of spacetime, we may appreciate time as the conveyor of a grand and still-unfinished drama of awakening. Even while the universe is running out of workable energy and dying a slow death thermodynamically, the passing of time allows the cosmos to be a momentous epic whose lengthy duration only enhances the suspense essential to a dramatic production. And even if it turns out that the awakening to life, sentience, and consciousness is confined to one middle-sized planet in an average-sized galaxy, it is nonetheless the entire universe that is undergoing that dramatic transformation locally.

The first twenty-one volumes in our thirty-volume set are devoid of life and mind. This is puzzling to an impatient, eternity-obsessed, design-fixated mind. The twenty-one or so lifeless volumes that stretch slumberingly from cosmic beginnings to the arousing of life in volume 22 look wasteful to earthlings who love magic. To those who are willing to wait, however, this extravagant span gives dramatic depth to the universe. During all the time prior to the emergence of life, the universe was not devoid of meaning. It was in the throes of a predawn awakening throughout.

Eventually, at the bottom of page 450 of volume 30, in a unique species of life, the cosmos begins to awaken consciously to meaning, goodness, beauty, and truth—that is, to an imperishable rightness that Abrahamic faith-traditions perceived to be rising on the horizon of the future. Only a nighttime watchfulness and patient expectation, not the magic-minded demand for perfect design, is prepared to read through our thirty volumes for a possible overall meaning. Again, only a dramatic, not a geometric, kind of inquiry could ever discern whatever this meaning might be. The search for meaning is by definition a search for a principle of comprehensive unity. The search for cosmic meaning is intent on finding a unifying principle that would integrate the totality of events into a rich narrative coherence.

Longing for magic, on the other hand, entails an impatience that the Abrahamic religious traditions consider unworthy of authentic human existence. The demand for divine creative magic, upon reflection, seems equivalent to insisting that the universe remain storyless and not subject to dramatic transformation over time. Neither otherworldly religiosity nor secular scientism can appreciate the deeply dramatic quality of nature's awakening. Devotees of the anti-Darwinian intelligent-design movement, for example, restrict the religious meaning of nature to its seemingly miraculous display of life's architectural intricacies. They scarcely notice the universe's long dramatic metamorphosis.[12] Some of their opponents, meanwhile, are also looking for evidence of perfect design in nature. Not finding it, they reject the idea of God, since they too assume that a divine creator, if there is one, would surely have made the world perfect immediately in the beginning.

Implications of Time-Denial

Both classical theology and modern scientific materialism have devised ways of ignoring the narrative quality of nature. Followers of both ways of looking at the world wonder why the universe requires an almost endless span of duration to become fully real and meaningful. Both biblical literalists and their materialist opponents cannot understand why, if God exists, the universe is not presently a fait accompli. If there exists an all-powerful creator, why is the world not a display of instantaneous and impeccable engineering rather than a long struggle toward the light?

Magic, I submit, may arouse our curiosity, but it cannot transform or renew our lives. Transformation takes time, and deep time allows for extra-long journeys of dramatic transformation. Both the spiritual conversion of persons and the developmental stages in cosmic history are transformations that take time. Deep time, then, is what keeps the universe from being a trivial display of magic. Neither traditional theologies nor contemporary scientific materialism, however, are predisposed to let the universe take its time. For many, including Einstein, the passage of time is not even real.

Because our individual searches for meaning take time, the seriousness of our personal quests is augmented by our seamless connection to the longer and larger cosmic vigil. Modern theology prior to Darwin and Einstein had not conceived of the universe as waiting for anything. Nor did it think of God—who was understood Platonically as the "eternal now"—as waiting. Yet in biblical traditions there has always been the motif of divine patience, a metaphor that makes no sense apart from waiting and accepting the passage of time as real and not illusory. The cosmic awakening, it would appear, also takes time, perhaps an immeasurable amount of it. And if time requires patience, then deep time requires profounder patience—indeed, the patience of God.

In any case, the demand for magic reflects a human impatience that blunts the raw edge of drama, narrows our temporal horizons, and robs us of the spiritual significance of having to wait for whatever has real value.[13] To souls that are bent toward magic, a much shorter span of natural history—and perhaps even no span at all—would be ideal. But to those who look for dramatic intelligibility, waiting is the better part of faith, of character, and of wisdom.

The metaphorical shift from "nature as design" to "nature as drama" invites us, then, to consider the possibility that the passage of cosmic time may be read without contradiction at different levels of meaning. If so, science and theology may be understood as distinct but compatible ways of reading a single, still-unfinished story of cosmic awakening. Furthermore, as soon as we realize that the universe is a temporal process, we no longer need to think of God's transcendence spatially—as up there or out there. If the universe is still coming into being, and if time is real, then transcendence refers also to what is up ahead, to what is not-yet.

The world thus leans not on the past, as the materialist assumes, but on the future, as hope requires.[14] The cosmos finds its stability—what Pierre Teilhard de Chardin (1881–1955) calls its "consistence"—not so much by losing itself in an eternal present in accordance with prescientific theology as by restlessly turning toward what is coming into time from up ahead.[15] True faith, at least in an Abrahamic setting, means being open to future possibilities that have not yet been actualized either in the cosmic past or in an eternal present. Faith is a disposition that lives off the sense of an unsettled future. The full awakening to rightness takes time, and true faith is willing to wait. Faith waits for what is possible to become probable, and then it continues to wait for what is probable to become actual.

Summary

For centuries the deep-time quality of nature was mostly hidden. The heavens seemed fixed forever in timeless splendor above. In the prescientific past people could easily assume that time was not real. Learning to embrace irreversible time has not come easily for either science or religion. Modern science itself emerged from an intellectual and religious setting in which the ancient Parmenidean and Pythagorean cults of timelessness, calcified by centuries of Platonic religious thought, still lived on in the minds of natural philosophers. Finding intelligibility, according to the habits of early modern science, still meant discovering timeless geometric patterns in nature rather than becoming aware of dramatic transformations going on in time. Philosophers had focused on the music of the spheres rather than on the epochal movements of matter. The main drama worth attending to was that of a person's soul seeking enlightenment by liberation from matter and time. After Einstein, however, we may at last envisage the whole universe as the primary drama and thus locate our personal spiritual pilgrimages, and even our great religious traditions, inside the larger narrative of cosmic awakening in real time.

Christian theology, in spite of its lingering love affair with Greek philosophy and its suspicion of time, is commanded by its deeper association with Abrahamic faith to view the whole of reality as historical and dramatic rather than finished and unchanging. To ancient and medieval theologians, the things of nature seemed to be imperfect copies of an eternal set of originals existing in timeless heavenly splendor apart from the physical universe. As a result, theology, in its search for perfection, lost touch with its Abrahamic roots and the anticipatory sense of time. It generally stripped nature of its narrative leaning toward the future of time. For centuries, theology failed to feel the pulse of the biblical hope for the fulfillment of all creation.

After Einstein, however, it is getting harder to separate our religious and personal quests for meaning from the fundamentally dramatic orientation of the whole universe that gave us birth. Einstein's portrait of the universe—corrected and enhanced by the contributions of Georges Lemaître, Edwin Hubble, Stephen Hawking (1942–2018), and countless others—is sufficiently interesting to bring fresh theological attention to the question of the meaning of time, cosmic destiny, and the significance of our own lives in time. In conversations on the topic of science and religion, the primary theological concern after Einstein, in my opinion, is not how to reconcile the raggedness of Darwinian evolution with divine providence but how to make sense of the whole cosmic story in which the evolution of life and the history of humanity are recent episodes. The metaphorical shift from nature as design to nature as drama invites us now to consider the possibility that the cosmic narrative, like other stories, is open to being read in multiple ways.

FOUR

Mystery

> The most beautiful thing we can experience is the mysterious.
> —Albert Einstein

> A person who is religiously enlightened appears to me to be one who has, to the best of his ability, liberated himself from the fetters of his selfish desires and is preoccupied with thoughts, feelings, and aspirations to which he clings because of their superpersonal value.
> —Albert Einstein

RELIGIONS TAKE FOR GRANTED the reality of mystery. Mystery is a name they give to what is most real but also out of reach. Cultivating a sense of incomprehensible mystery is an exercise encouraged by almost all of the world's spiritual traditions. Mystery, they say, can seize and ennoble us, but it remains beyond our grasp. Science, on the other hand, is eager to comprehend the world. Nothing, it seems, lies beyond its power to penetrate. Sometimes we become so impressed by science's capacity to comprehend things that we think its point is to eliminate mystery.[1]

Einstein did not agree. Science, he thought, should expand not only our understanding of nature but also our sense of mystery. "The most beautiful experience we can have is of the mysterious," he exclaimed. "Whoever does not know it and can no longer wonder, no longer marvel, is as good as dead." He adds: "It is this knowledge and this emotion that constitute true religiosity."[2]

For Einstein, mystery was real, not an illusion or a cover-up for current scientific ignorance. The awareness of mystery should therefore grow, he thought, rather than shrink as scientific research becomes more successful. The greatest mystery of all, Einstein claimed, is that the universe is comprehensible.³ Mystery, for him, was a mostly hidden source of meaning that gradually revealed itself—especially through geometry—to minds properly prepared. Is it not remarkable, we might ask with Einstein, that the whole of nature is permeated by an unseen intelligibility that invites us to explore it without end? Without the great mystery of the universe's intelligibility, human inquiry, including scientific research, would be inconceivable. Nor would thought exist or questions be asked. The best way to treat our minds is to keep alive a sense of mystery.

Mystery, then, is much more than a space to be filled in by our own intellectual accomplishments, and it is not the same thing as a problem. A problem is a puzzle that can eventually be solved by human ingenuity. Mystery, however, cannot be solved. Instead of disappearing incrementally as science advances, mystery keeps looming larger than ever. It is like a moving horizon that retreats and expands as scientific understanding advances. It does not go away, nor does progress in human understanding ever lessen it. Mystery is indestructible. No wonder people sometimes call it God.⁴

Einstein thought of religion as a response to mystery. He referred to his own attraction to mystery as a cosmic religious feeling. By turning our minds toward mystery's inexhaustibility, true religion in effect invigorates science by providing assurance that research will never come to an end. It would never have occurred to Einstein that mystery is a fleeting human invention or that it can ever be whittled down by scientific progress. As I have already pointed out, Einstein was in love with eternity more than with time. He even longed to attribute indestructibility to the physical universe. As it turns out, he was wrong to eternalize nature, but he was not wrong to assume that mystery is imperishable. This brilliant thinker, the greatest scientist of our age, was convinced that the love of mystery is essential to the spirit of exploration. He pitied any scientist who lacked this sensibility. His appreciation of mystery, in my opinion, aligns him more closely with the religious majority of his fellow humans than with the minority of scientific skeptics who claim him these days as their champion. Like most religious believers, Einstein did not assume that reality consists solely of what our senses can perceive or what science can know. In bowing to mystery, he acknowledged that there is infinitely more to reality than the human mind can ever encompass.

In Christian theology as well, the term "mystery" designates much more than a gap in our knowledge or a vacuum to be filled in by growth in human understanding. Mystery is another name for God.[5] Mystery may be thought of in many ways, but I understand mystery especially as the indestructible horizon of the not-yet. To Christian faith, mystery includes more than a divine presence beneath or above the plane of what we can sense. Mystery is also the inexhaustible future that is full of possibilities not yet actualized. So, to suggest again a term I introduced before, a good name for the divine mystery is Absolute Future.[6] In keeping with the central thrust of Abrahamic religion, mystery may be understood not as an eternal present that rescues us from the flow of time but as the inexhaustible future into which all of time streams. Christian faith expects mystery to reveal itself not so much in nature's hidden geometric order as in a dramatic synthesis yet to be finalized, in a coherence for which we have to wait—ideally in a spirit of active and attentive hope.

Is Mystery Personal?

Aside from the sense of mystery, however, Einstein considered everything in religion to be mere superstition, including the idea of a personal God. Even though Einstein sometimes talked about God, he rejected the notion of a creator who loves, judges, and cares. To be credentialed as a deeply religious person it was enough, he thought, that he had a ceaseless appreciation of mystery. In fact, it was owing to his profound sense of mystery that he dismissed belief in a personal God as a false and deficient kind of piety. The whole idea of a personal God, he claimed, was a palliative that humanity should by now have gotten over.[7]

Why the universe is comprehensible at all Einstein could not say. He let that mystery remain incomprehensible. An appreciation of the universe's comprehensibility is indispensable to the mindset of any true scientist, he insisted, but why the universe is intelligible in the first place is not a problem that science can ever solve. Einstein attributed the very possibility of scientific understanding to the gracious existence of an indestructible cosmic intelligence, but this intelligence was, he thought, impersonal. Here again we can detect the ghost of Spinoza. Einstein agreed that his own mind, as well as that of other humans, could function only because a cosmic intelligence had gifted the universe with a meaning that was decipherable through pure mathematics. That cosmic intelligence, according to Einstein, was impersonal.

How, though, we may ask, can humans fully respect a mystery that does not itself rise to the level of personhood? If mystery is less than personal, is it not, in a sense, smaller than we are? In human experience, after all, nothing has more intensity of being than the persons we encounter. Is it not in the words and faces of other persons that mystery, even while hiding, breaks through most powerfully to each of us? So how can the mystery revealing itself in nature receive full reverence from us if it is ultimately impersonal, if it is an It rather than a Thou? These, at any rate, are questions that Christian theology is obliged to ask in its conversation with Einstein.[8]

In intellectual culture today, I do not need to point out, the universe seems not only impersonal but sometimes devoid of mystery. Scientifically educated people often assume that a sense of mystery, along with ancestral myths that pointed to it, is nothing more than a childish illusion. Religions and myths are fictions that make the purely physical universe seem personal, but they are fabrications that reasonable people should have outgrown by now. To the rationalists and the scientific debunkers, even Einstein's cosmic religious sense is just a vague feeling, not an indication of something incorruptible. For some scientists today, the sense of mystery has vanished altogether. As a consequence of science, says the late physicist Heinz Pagels, "the universe will hold no more mystery for those who choose to understand it than the existence of the sun." Accordingly, "as knowledge of our universe matures, that ancient awestruck feeling of wonder at its size and duration seems inappropriate, a sensibility left over from an earlier age."[9]

Einstein, by contrast, held that mystery does not dwindle and vanish. Science can neither exist nor make progress unless the universe's intelligibility is limitless. Thoughtful scientists trust that beneath the surface of first impressions and scientific abstractions lies an inexhaustible depth that invites endless exploration. This depth is not incomprehensible but infinitely comprehensible. Moreover, conscientious scientists confess that they need infinite horizons not just to account for continual growth in scientific understanding but also to give zest to their personal lives. Most of us, Einstein would agree, are not satisfied simply to eat, mate, work, and sleep. We want more, indeed infinitely more. Unlike our fellow animate beings, we cannot live and thrive without a sense of belonging to a mystery unfathomably larger than the sensible world. Indeed, we cannot remain fully human without a sense of this mystery. Both Einstein and theologians can agree on this. For Christian faith, however, mystery cannot be less than personal if it is to be an ultimately satisfying source of meaning as well as a liberating challenge to personal beings.

Immensity and Theology

Mystery, to be sure, does have nonpersonal—or superpersonal—attributes. For example, an awareness of mystery, whether in the case of Einstein or in the case of theology, requires that we first have a sense of immensity. In this respect Einstein's physics—much more than his religious opinions—is theologically significant. His science has extended not only our sense of time's depth, as we observed in the preceding chapter, but also our impressions of spatial magnitude. To feel mystery's inexhaustibility, it seems essential that we have the sensory experience of an indefinite spatial openness. Genuine religion requires the extension of the mind toward what is great—*extensio animi ad magna*—as medieval mysticism understood. We need to stretch our minds and imaginations toward sheer vastness to avoid feeling suffocated. The feeling of being confined makes us anxious, whereas a sense of immensity brings release from limits that oppress us. Happily, Einstein's science, much more than his religious opinions, has blown off the lid.

God, says the psalmist, has put us in a "wide setting where there is no sense of being hemmed in."[10] For centuries the sweep of the heavenly spheres in perfectly circular orbits enlarged and ventilated the souls of our ancestors. But in the early days of modern science, after Tycho Brahe, Johannes Kepler, and Galileo Galilei had exposed some of the blemishes in the superlunary spheres, the heavens began to seem ordinary. The modern scientific search for mathematical simplicity brought the stars down to Earth. And even though post-Copernican astronomy, Darwinian biology, and Big Bang cosmology have made the natural world look large and alluring, scientific materialism, by reducing the world to inert physical bits, now threatens to make it look smaller than ever.

Classical theology, to its credit, insists on the infinity of mystery, so there has never been a reasonable theological excuse for our impoverished images of deity. Infinity is unsurpassably larger, is it not, than any imaginable cosmic immensity? So, God, if God means infinite mystery, should never have become too small to inspire worship. Even Einstein's own thought-world was at times dragged down by his attraction to materialist atomism, just as it was hemmed in by his initial inability to imagine the immense depth of time's passage. Nevertheless, his general theory of relativity, complemented by a century of new discoveries in astronomy and astrophysics, has now revealed a cosmic spatial magnitude that has exceeded everything previously imaginable in the realm of extension.

To get just a glimpse of the observable universe's colossal spatial range, start by considering the diameter of the Milky Way, an average-sized island in an endless ocean of constellations. If you sent a message from one edge of our galaxy to an imaginary recipient at the opposite edge at the rate of the speed of light—186,000 miles per second—it would take 100,000 years for your greeting to arrive at its target on the other side, and another 100,000 years to get a reply. Recall now that it was during a span of 200,000 years that modern humans gradually emerged from ancestral tribes in Africa and spread out over the face of the earth. A generous portion of human history would fit into the time period during which you are awaiting a response.[11]

As we journey toward immensity, we should keep in mind that ours is just one galaxy. Billions more are out there, some full of stars orbited by planets that may be hosting living complexity or something like it.[12] Furthermore, cosmologists are now speculating with increasing confidence—so far unanchored to any experimental data—that countless undetectable universes exist outside our own. If these cosmic speculations are right, and they may be, the Big Bang universe is only an infinitesimal part of a much more colossal immensity known as the multiverse.

For countless educated people, science rather than theology widens the world and opens the eye of contemplation. Theology, they suspect, has failed to expand its sense of God in proportion to cosmology's unprecedented new awareness of spatial and temporal enormity. Science has caused our minds to outgrow earlier depictions of deity, whereas theology has yet to feel fully the recent scientific magnification of nature. The idea of God, at least for many educated people, now seems too small by comparison to arouse either intellectual interest or the sentiment of worship. Today, I believe, the high cultural disillusionment with conventional religion stems in great measure from the failure of theologians and religious teachers to give us a God larger than the universe.

Religious education in homes, schools, colleges, and churches is complicit in this downsizing of deity. Many of us picked up our sense of God in Sunday school, long before we became adults. Our earliest images of God became fixed in our sensibilities at a time in our lives when we knew little about the natural world. As we grew up, our minds expanded, but our images of God failed to keep pace and became too small to arouse a profound sense of wonder. Our deities disappeared beneath the surface of new portraits of spacetime. Leaving the cosmos to science, modern Christian spirituality then became so privatized at times that it ignored the new discoveries of spatial and temporal expansiveness, assuming that immensity was irrelevant to the religious quest for meaning.

The Mystery of Superabundance

Given the nature of human consciousness, we cannot think of God without the use of images. So if our imaginations are restrained, our sense of God becomes too small for our souls. Jesus, the Gospels tell us, found that the chief stumbling block to accepting his message, was smallness of faith, whether that of his disciples or that of the crowds that came to hear him.[13] To electrify their devotion he had to stretch their imaginations, which he undertook to do by framing his preaching in shocking pictures of divine excess. In announcing the coming of the Kingdom of God, he portrayed the mystery of true rightness as unreasonable extravagance. In both his practice and his preaching he presented images of an immensely wasteful deity. Indestructible rightness, at least for Jesus, came to expression in portraits of divine superabundance.

Jesus's own religious life was attuned to the depictions of limitless divine generosity that he had picked up from the traditions of his ancestors. As a devout Jew, he prayed to a God who spreads out a banquet for us in the sight of our foes and makes our cups to overflow. His God is one whose anger always gives way to mercy and whose "rightness" (*sedeqah, mishpat*) pours forth like a waterfall.[14] Jesus had learned from the Torah, the Psalms, and the Prophets that the mystery of true rightness can be conveyed only in images of overflowing. His own boundary-breaking hospitality to sinners and outcasts gave life to his thoughts about the immoderate inclusiveness of God. Consequently, his social life, table fellowship, and public preaching offended those who wanted a God small enough to be simply fair in the distribution of gifts. The graciousness of God embraces sinners and saints alike, Jesus emphasized, and it does so out of all proportion to what they deserve. Divine rightness tears down the stiff borders of conventional reasonableness, with its tidy assumptions about ethical fairness. The strange God whom Jesus called Father allows rain to fall and the sun to shine on the just and the unjust alike. A large faith, in that case, must reflect the divine overindulgence.[15]

Science's recent disclosure of spatiotemporal immensity almost physically links our imaginations to boundlessness. Flooded by new astronomical impressions of nature's depth and breadth, our religious imaginations may now indulge in an unprecedented magnifying of mystery. Contemplation of cosmic immensity may, if we let it, finally liberate our sense of God from its all-too-terrestrial captivity.

In contrast to Jesus's God of excess, there is the conventional god of equivalence.[16] This dull deity is the product of a small faith that settles for balance in place of boundlessness. The god of equivalence is a small-minded appraiser whose logic is riveted on fairness and who demands that the right price be paid for everything. Unlike Jesus's God of excess, the conventional god of equivalence demands that every fault receive a proportionate punishment; that the two sides of a scale always rise or sink to the same level; that there is no free lunch; that everything that goes around comes around; that life is a zero-sum game; that laborers in a vineyard, to use Jesus's parable, be compensated in exact proportion to the amount of time spent working.[17]

Jesus's God does not function in this transactional way. God is a mystery whose identifying trait is not fairness but excess. The God of superabundance shatters the standards we set up in our minds, institutions, schools, workplaces, moral systems, and theologies to make us feel righteous. Jesus's God breaks through the barriers we erect to protect our narrow and exclusivist criteria of self-worth. His God is unimpressed by the praise we bestow on ourselves for being good or the fear we have of not measuring up. This is why Jesus idealized little children—those not yet fully socialized—since they carry no resume on which to rest their self-esteem. It is the unfettered minds of little children that open us to the boundlessness of mystery.[18]

Jesus's preference for the logic of superabundance is bound to affect how we treat others: "You have heard that it was said, 'An eye for an eye and a tooth for a tooth.' But I say to you, do not resist an evildoer. But if anyone strikes you on the right cheek, turn the other also; and if anyone wants to sue you and take your coat, give your cloak as well; and if anyone forces you to go one mile, go also the second mile." Again: "You have heard that it was said, 'You shall love your neighbor and hate your enemy.' But I say to you, love your enemies and pray for those who persecute you."[19]

It is soul-expanding to watch how the logic of superabundance works. God's compassion embraces those who have strayed and extends mercy to them long before they have returned to the fold. Forgiveness, in other words, is granted before confession.[20] Being born blind is not punishment for sins but an occasion for God to let in the light of a rightness opposed to the ironclad morbidity of equivalence.[21] When Christians confess in the

Nicene Creed that they "believe in one God, the Father almighty"—that is, in what Jesus called *abba*—they are professing their devotion to the higher rightness of superabundance, not the lesser good of equivalence, which leads to self-hatred, scapegoating and endless expiation.

Here I am following the philosopher Paul Ricoeur's illuminating interpretation of Jesus's teaching.[22] Ricoeur makes a crisp distinction between the conventional "logic of equivalence" and Jesus's "logic of superabundance," a distinction that proves helpful in our own attempts to make theological sense of the universe after Einstein. For example, fixation on the word "law," employed by scientists to account for nature's predictability, sometimes unconsciously reflects and legitimates the juridical ideal of equivalence. As we shall see, the idea of lawfulness is not enough to capture either the character of God or what is really going on in the universe. We meet up with the mystery of God not so much by looking at the inflexible laws and mathematical consistency that govern nature as by following the indeterminate cosmic story whose full intelligibility overflows geometric understanding and belongs to a future that has not yet come fully into view.

In spite of Einstein's personal preference for a universe locked into a timeless geometry, his own calculations, once the dimension of time is taken to be irreversible, have liberated the cosmos from the box of eternal necessity in which modern scientific thought had imprisoned it. Nature's intelligibility has turned out to be dramatic and not just geometric. Einstein was fixated on the geometry rather than the drama. For this architect of contemporary cosmology the dimension of mystery lay mainly in the harmony of nature, not in the indeterminacy of an unfinished story. For theology after Einstein, the mystery of the universe consists mainly of the far-from-finished drama of awakening that is still weaving itself into the fixed loom of nature's regularities.

As the carrier of a long and still-unfinished story, the cosmos has now burst out of the limits imposed on it by pure geometry. For the past four centuries, philosophers have highlighted nature's mathematical consistency more than its uneven narrative openness. Einstein himself never acknowledged that the universe could be carrying a kind of intelligibility that cannot be captured by mathematics. His mentor Baruch Spinoza, under the influence of René Descartes, had so idealized geometry that he even wrote his main work, *The Ethics,* by organizing its postulates in what he called a "geometric manner" (*more geometrico*). In the early twentieth century a widely shared geometric trance was still preventing Einstein from allowing that anything truly new could ever happen in the universe. Even while breathing in the fresh air of mystery, Einstein formally claimed that everything that happens is already determined by immutable physical laws arranged according to strict geometric principles. In his ideal universe, the temporal passage essential to narrative meaning was an illusion in which nothing inherently surprising could ever really occur.

Yet it was a careful interpretation of Einstein's own theory of general relativity that let loose the good news of nature's dramatic capacity to transcend necessity as it awakens to the mystery of the not-yet. In spite of Einstein's uncompromising intellectual commitment to the ideal of a cosmos bound by lawful constraints, his own science leaves ample room, as others have shown, for a long cosmic drama that is still open to unrealized possibilities. And if the universe is a drama still going on—as time passes, space expands, and matter becomes more complex—it is not at all self-evident that its future will be simply a return to, or the equivalent of, what has always been.

A culturally entrenched logic of equivalence conceals from scientific thinkers the narrative quality of their depictions of nature. Even if the cosmos is destined to revert physically to primordial quiescence during some future epoch of energy collapse, it will have given birth to a narrative content that was never fully predictable. The universe's dramatic excess liberates it from sheer necessity, even though no laws are broken in the process. It is not in conformity to inviolable regulations but in temporal openness to new narrative patterns that the universe exceeds the bounds of balance and equivalence.

Along with Spinoza, Einstein idealized a changeless universe. Countless other scientists and philosophers have also settled for balancing equations instead of appreciating the superabundance of nature's narrative digressions—such as the birth of life and the flourishing of thought. Even after Darwin, scientific materialists have interpreted life's wild wanderings, not as evidence that nature is an uncertain drama, but as the pointless outcome of physical determinism and natural selection. The Darwinian philosopher Daniel Dennett, for example, explains away the whole drama of life as the mere algorithmic reshuffling of simple elements by impersonal physical laws that have always been around.[23]

As far as scientific method is concerned, an explanatory appeal to the invariant rules of nature is unproblematic. Scientific method does not expect to find, nor does it look for, a universe open to the novelty of dramatic excess. As a scientist, Einstein rightly focused on what is constant in nature, symbolized for him by the unchanging speed of light that boxes nature in geometrically. Because of his obsession with equivalence in nature, however, Einstein's philosophical reflections on nature overlooked the irreversibility of time and the universe's indeterminately dramatic way of being and becoming.

Like many other modern thinkers, Einstein was more impressed with the universe's apparent determinism than with its wild journey of awakening. His ideal universe was one that could not deviate in any way from necessity. Modern scientific thought in general has followed the same legalistic logic, one that settles for predictability and ignores the turbulent drama of cosmic awakening. What enchants scientific thinkers throughout the modern age is the lesser rightness of equivalence, not the higher rightness of excess. Social and political ideologies likewise have generally surrendered to the ideal of equivalence—sometimes calling it justice—whose maintenance demands the imposing of penalties to redress imbalances. In the natural sciences the phenomena of evolution, complexity, and chaos seem congenial to a narrative reading of nature, but scientists instinctively assume that even these must submit finally to the logic of equivalence underlying materialist determinism.

Scientific simplification irons flat the wrinkles in cosmic process that have opened the universe to dramatic uncertainty throughout its long history. But while equivalence rules the universe mathematically, superabundance rules it dramatically. The modern ideal of mathematical simplification is scientifically useful, but it can divert our attention from the fact that something new and unrehearsed is always beginning to take shape up ahead in the cosmic story. The modern materialist ideal of intelligibility-as-simplification leads scientific thought to overlook the novelty that may be barely blossoming in the immensity of deep time and the expansiveness of space. The mathematical simplification essential to science, if taken as fully representative of the concrete world, steers our imaginations away from the greatest mystery of all—the dawning of the not-yet.

Classical theology—I am thinking especially of Saint Bonaventure (1221–1274)—sometimes revived the biblical images of divine overflowing and the excess that Jesus associated with God. Most of the time, however, theology has fostered an impression that the fullness of being has already been actualized in eternity and that what happens in time contributes little or nothing to the primordial fullness of being. So, not only modern materialism but also traditional theology hides from the prospect of a genuinely new future for the universe. Today, neither theology nor deterministic science allows sufficient room for the not-yet in which "all things are possible."[24]

In reciting the Nicene Creed after Einstein, Christians may now acknowledge not only that God has entered into time but that the theological meaning of time is to reveal the mystery of divine superabundance. By professing their faith in the incarnation of the Son of God, Christians are encouraged, not to leave the temporal world behind or to turn their eyes away from it, but to align themselves fully with its awakening to the mystery of the not-yet.

Summary

In Christian theology, mystery is another name for God. The newly discovered wasteful expansiveness of time and space allows us to read the universe as a radiant expression of divine superabundance. Mystery has always hovered at least vaguely at the edges of human awareness, but some modern scientists and philosophers think the objective of science is to wipe it away. Einstein, on the other hand, thought that science can expand our sense of mystery. He did not assume, as other scientists have, that reality consists solely of what our senses can perceive or what science can know. In bowing to mystery, he presupposed that there is infinitely more to the real world than the human mind can ever encompass.

In spite of his exceptional appreciation of the mystery of matter's comprehensibility, Einstein clung to the modern materialist and determinist philosophy that has generally had the effect of marginalizing mystery and God. The impersonal laws of physics are timeless and effective enough, Einstein agreed, to govern the universe without divine providence. So, after Einstein, we still have the question What place is there for God, if any? Today, much scientific thought continues to share Einstein's assumption that everything that happens in cosmic history is simply the impersonal, inertial uncoiling of what was lawfully laid down at the beginning. Nothing, in that understanding, can escape the invariant rules of physics. Indeed, everything that happens is determined by implacable regulations that leave no space for divine influence or human freedom. One response to this common interpretation is to look for mystery not so much in the inviolability of nature's invariant habits as in in the extravagant drama of a cosmic awakening.

FIVE

Meaning

> The scientist is possessed by a sense of universal causation.... His religious feeling takes the form of a rapturous amazement at the harmony of natural law, which reveals an intelligence of such superiority that, compared with it, all the systematic thinking and acting of human beings is an utterly insignificant reflection.... It is beyond question closely akin to that which has possessed the religious geniuses of all ages.
> —Albert Einstein

> At the basis of the whole modern view of the world lies the illusion that the so-called laws of nature are the explanations of natural phenomena.
> —Ludwig Wittgenstein

> The future enters into us, in order to transform itself in us, long before it happens.
> —Rainer Maria Rilke

ALBERT EINSTEIN THOUGHT OF himself as a person of faith. As we have already seen, he was in love with eternity. He was also proudly devoted to the indestructible superpersonal values of truth, goodness, and beauty. By adhering to these timeless values, Einstein believed, a person may be liberated from selfish desires—a deliverance that the main religious traditions also prize. Moreover, for Einstein, devotion to timeless values was essential to doing science rightly. He understood that the search for scientific objectivity requires a devout passion for truth, a resolve that is not merely ethical but also religious. Additionally, as we have seen, Einstein was something of a mystic. He would not have called himself one, but he confessed to being wonderstruck by the mystery of the universe's lawfulness and comprehensibility. Clearly, this commitment gave his life meaning and joy.

Although Einstein was Jewish by birth, he was not traditionally religious and did not believe in the God of Abraham. Occasionally he talked about God, whom he also called the Good Lord or the Old One, but he was no conventional theist. He had been exposed briefly to Catholicism in his earliest years, but he could not accept the idea that a personal redeeming God exists distinct from nature. The biblical God who creates, responds to prayers, and allegedly intervenes in the world was, to him, both morally and intellectually indefensible. In fact, Einstein considered the idea of a personal God the main obstacle to reconciling science with religion.[1]

Why so? Because a personal, interested God, one who can answer prayers and heal suffering, would have to intervene in nature to make things right. In doing so, divine action, Einstein reasoned, would have to violate the eternal laws of nature essential to scientific understanding. A personal God, to be known as such, would have to suspend the laws of physics upon which modern science bases its confidence in making predictions. And if the rules of nature could be set aside by supernatural manipulation, even if only occasionally, such disruptions would undermine science's predictive credibility, exposing the habits of nature as provisional and temporary rather than eternally valid.

The sense of his own life's meaning was inseparable from Einstein's scientific assumption that the laws of nature are timeless and inviolable. In his view, a life dedicated to preserving the integrity of science left no room for belief in the existence of a personal, interventionist deity. It is important to note, however, that Einstein rejected the idea of a personal God not out of a commitment to atheism, from which he deliberately distanced himself, but for the sake of what he took to be authentic religion. In addition to affirming superpersonal values, true religion, for Einstein, meant surrendering humbly to the great mystery underlying the mathematical harmony of the physical universe. Surrendering to nature's exquisite order was essential to the whole scientific enterprise, whereas belief in the existence of a caring and active God would amount to a wholesale repudiation of the scientific worldview. The scientist may talk about God as long as that name is taken to mean the timeless impersonal intelligibility revealed in science's discovery of nature's wondrous mathematical consistency.[2] But God, in that case, is not the author of life, nor the liberator of Israel, nor the faithful one who makes promises and sends prophets into our midst. Nor is God, in that case, the compassionate power that created the world out of love.

And yet, as we have seen, Einstein was passionately in love with eternity, a theological attribute that he ached to identify with nature itself. For a while he devoutly believed that, except for local fluctuations, the universe was fixed forever in a state of virtual completeness. The noblest task of science, given that understanding, was to make nature's timeless hidden structures transparent to the human mind here and now. Participating in this revelatory task was as much an act of religious devotion as an adventure of scientific discovery.

Reading the Universe

Because Einstein was so attached to eternity, he failed to find any religious meaning in time. Although he did not realize it at first, his geometry had in principle left room for the universe to take shape as a large story still going on in time. If nature's fundamental format is that of a story, theology after Einstein does not ask whether nature points directly to a First Cause or an Intelligent Designer. Rather, it asks whether the story carries a meaning that has yet to reveal itself. In addition to asking with Einstein whether the universe has a geometric intelligibility, theology may also ask whether it has a narrative, or dramatic, intelligibility unknown to science and not yet fully actualized in time.

But how and where would theology look for this dramatic meaning? Not in exceptional ruptures of the causal continuum or in the physical fabric of nature, as Einstein's caricature of theology supposes. Contrary to Einstein's personal opinions, theology does not have to look for miraculous exceptions to the rules of nature upon which science is based. Instead, the task of theology is that of looking into the universe for a narrative meaning that somehow grasps hold of us without violating any laws of nature. Presently, if the universe is still emerging into being, any such meaning is partly, if not mostly, out of reach. If it is to be found, or if it is to find us, we must turn patiently toward the region of the not-yet. Epistemologically speaking, we would have to adopt what I shall be calling an *anticipatory* stance in our reading of the universe. To encounter cosmic meaning we would have to wait for it, not passively but patiently and attentively.

Let me explain. The universe is not a design but a drama, a claim that I will later develop more precisely. Here I will just say that the kind of meaning we look for in a dramatic universe is analogous to what we expect when we are reading a novel, attending a play, or watching a movie. In these media we seek dramatic coherence, not mathematical clarity. So, if the universe is a dramatic coming-into-being, we cannot allow ourselves to be too impatient in our pursuit of its meaning. Narratively speaking, it is unrealistic to expect that a novel's main point will become fully transparent in the first few chapters. In watching a film or attending a play, we cannot reasonably demand that the first several scenes will reveal what it is all about. To find out what is going on in any narrative production we have to curb the desire to have everything make sense immediately. Patience is also an epistemological prerequisite in the human search for meaning. In terms of our new cosmic story it is the function of faith to nurture such patience, to give us the courage to watch vigilantly for a dramatic coherence that may not yet be fully available.

A dramatic universe is also presupposed by the Darwinian understanding of life. To know life, Darwin had to tell a story about it. For any series of events to be a story, it has to be a mixture of three ingredients: unpredictability, consistency, and temporal duration. First, without some degree of unpredictability, any series of events would be completely deterministic. Without accidents or contingency, in other words, there would be no element of surprise and suspense. Second, without some degree of predictability or consistency, the universe would be too unsteady to be a story or to host the drama of life's evolution and the arrival of thought. The element of consistency is supplied by the unbending habits of nature that we refer to as laws. Third, to have a story, the span of time must be long enough for a series of events to carry a meaning.

Long before the arrival of the first living cells on Earth, the universe had acquired a remarkable disposition to host stories such as the story of life. The physical or cosmic conditions required for the evolutionary story—contingency, predictability, and the passage of time—were waiting to spin bits of matter and threads of molecules into a drama of life as soon as the first cells arrived. Darwin's science cannot account for this threefold narrative recipe, since evolution presupposes it. The natural world, moreover, is still undergoing a transformation for which the word "drama" is more appropriate than a term like "design" or "plan." The theological significance of our focus on drama rather than design is that a drama can be the carrier of a meaning that presently lies hidden in the future.

If the passage of time holds meaning, then, we have to wait for it—not passively but patiently. Patience, of course, is a virtue fundamental to Abrahamic religions. After Einstein, theology in the Abrahamic traditions need not look for God either in exceptions to the normal flow of events or in elusive "causal joints," where God is expected to modify natural processes. Theology after Einstein is not obliged to look for the imprint of divine influence in subtle microphysical stirrings or in any physical events that might require a breakdown of nature's habitual and reliable routines. Instead, the primary place to look for God is in the mystery of the not-yet. Theology, as I understand it, looks for the presence of God not in the breaks but in the blossoming of nature.

Concealed beneath the mathematical formulas by which Einstein connects mass to space and time, we find a universe gradually revealing itself as a long and far-from-finished drama of awakening. The universe, then, is not a static, timeless design trapped in sameness from one age to the next. The universe is not a perfect architectural accomplishment. Rather, it is a suspenseful epic of matter coming to life, of life becoming alert, of alertness looking for meaning, and of faith awakening to indestructible rightness.

Before we found out recently from science that nature is a story still being told, the question What is really going on in the universe? could scarcely have arisen. Had we lived before Darwin and Einstein, we may have worried about what is going on in our souls or in human history but not about what is transpiring in the cosmos. We would have looked upon the physical world as a stage for the human drama or as a point of departure for our personal quests for meaning, freedom, and salvation. We would not have noticed that the universe itself is a grand drama of awakening. Along with the philosopher Immanuel Kant (1724–1804), we would have wondered perhaps about the moral law fixed in our hearts and about the starry skies above, and these may have pointed us toward God. But prior to the twentieth century we would not yet have heard the impressive scientific story that links us more tightly than ever to the stars, the galaxies, the periodic table, the sun, the planets, and life. Until very recently we could scarcely have realized how intricately life and thought are tied into sidereal movements and microphysical events going all the way back to the remotest epochs of the cosmic past.

Until the last century or so, human beings knew nothing about the long story that has spun our souls out of the stars. For centuries the physical universe seemed stationary to natural philosophers, and the physical universe was, for religious believers, primarily a place for working out their personal salvation. Human beings had no inkling that the whole cosmos is a suspenseful, transformative passage from mindlessness to thought, from abiding in darkness to longing for light. Nor would they have worried much about what happens to the universe after our souls have left it for the hereafter. Traditional theology's lack of interest in the cosmic journey is forgivable, of course, since only after Einstein could it have learned that the entire universe is a continuous narrative and not just a platform from which to launch our spiritual adventures. Until recently we did not know that nature—over an immense amount of time, though not without setbacks, wild wanderings, and dead ends—has become dramatically alive, sentient, conscious, and, more than occasionally, compassionate. Surely something significant is going on here.

Looking at nature narratively, as a drama of awakening, we have a whole new way of framing the conversation between science and theology. Today, I suggest, the main objective in this exchange is not to specify how God acts in nature but to learn how to read the story of the universe. I believe there are three main ways of reading the cosmic story: the way of *archaeonomy*, the way of *analogy*, and the way of *anticipation*. These readings correspond, respectively, to three ancient but persistent perspectives on the world, those of Democritus, Plato, and Abraham. Let us study them briefly now, and then inquire in later chapters—and in conversation with Einstein—how each of these three ways of reading the cosmos interprets the topics of origins, life, thought, freedom, faith, hope, compassion, and caring for nature.

Archaeonomy. The pre-Socratic Greek philosopher Democritus (ca. 460–370 BCE) thought that if you want to understand the cosmos, all you need to be aware of are, first, the infinite void of space and, second, the constant reshuffling of countless atoms within it. Underneath the outward complexity of nature and the turmoil of human affairs churns an endless combining and recombining of simple, eternal, indivisible bits of matter. Democritus and his followers referred to these irreducible units as atoms, a word that in Greek means "unable to be divided." At the foundation of all being, the atomist maintains, there exists nothing but an infinite number of mindless elemental particles moving around aimlessly, coming together in various patterns, falling apart, and then reassembling within the limitless expanse of space.

The simplicity of Democritean atomism is irresistible. Even today many scientific thinkers approach the physical universe in roughly the same way as Democritus did. After Einstein, however, atomism cannot mean exactly the same thing as in antiquity. Unlike the atomists of old, contemporary atomists now realize that the universe is not eternal and necessary but is still coming into being, producing unprecedented outcomes along the way. Consequently, atomism—breaking things down into their tiniest parts—now has the additional connotation of going back in time to the earliest period of cosmic history. Let us call the contemporary atomistic way of looking at the history of nature *archaeonomy*. The term is composed of two Greek words, *arche*, which means "origin" or "beginning," and *nomos* which means "law." The archaeonomic reading of nature implies that the state of the universe as it was in the beginning (arche) lays down the law (nomos) for everything that happens later on.

Every event in cosmic history, according to the archaeonomic stance, is the result of physical causes, conditions, and constants present from the start. To understand all the outcomes of the cosmic epic, our minds, by way of analytical science, have to travel figuratively all the way back to the beginning of cosmic time. Since we cannot literally journey back that far, we may partially recapture the opening cosmic scenery by breaking present things down into their simplest parts.

But is archaeonomy a reasonable and successful way of making sense of a still-awakening universe? By leading our minds back in time to the original preliving and mindless stage of natural history, the archaeonomic stance fails to make sense of the actual story that has taken place since the universe's first moments. Archaeonomy methodically ignores the dramatic nuance and unpredictable narrative twists and turns in the cosmic awakening. One of the outcomes of cosmic process is the recent arrival in natural history of the very minds through which science is now undertaking its retrospective journey in time. Archaeonomy fails to capture the dramatic depth of this great event in cosmic history.

Breaking complex things down analytically, I want to emphasize, is an appropriate method for scientists to use in their quest to understand the natural world. But archaeonomy is not science. It is a worldview based on the solemn belief that the only reliable way to understand the world around us is to trace everything back analytically to how things were in the beginning.[3] An exclusively archaeonomic dig into the cosmic past implies—metaphysically—that true being is virtually equivalent to what has happened already in the remotest past period of cosmic history. At bottom the universe is still just atoms and the void. In this archaeonomic way of reading the universe, what is yet to come is destined to be merely a reassembling, by the invariant laws of nature, of the same elemental units that were present in the beginning. The passage of time brings no new being or meaning.

An outspoken representative of archaeonomic atomism is the Oxford scientist Peter Atkins, who declares that all present phenomena, no matter how wondrous they may seem to be at first, are really nothing more than elemental atomic simplicity "masquerading as complexity."[4] I should point out that there is a generous dose of archaeonomy in Einstein's reading of nature as well.

No doubt, "archaeonomy" is a strange new word. Why not just call it materialism and leave it at that? I use the new term because after Einstein, scientific analysis means not only breaking complex things down into their simplest parts but also returning in time to the earliest stage of cosmic process. "Archaeonomy" adds a new temporal accent to the customary definition of materialism. It implies not only that complexity is reducible to simplicity but also that the present is nothing more than the past, now colored over with filmy present-day appearances. The purest state of being, archaeonomists declare, lies in the earliest physical stage of cosmic beginnings. Archaeonomy amounts then to a "metaphysics of the past," which we shall now contrast with the analogical "metaphysics of the eternal present" and the anticipatory "metaphysics of the future."

Analogy. We may read nature analogically instead of archaeonomically. That is, we may think of beings that exist in time as though they are imperfect copies of imperishable forms that exist outside of time. This influential way of reading nature is associated especially with the great apostle of timelessness, the philosopher Plato (ca. 428–348 BCE). Plato and his followers advise us that to grasp the world's intelligibility, to read nature rightly, we must look at objects and organisms as imperfect copies of perfect but invisible heavenly counterparts. I call this way of reading the universe analogical because it holds that natural beings are analogous to eternal, ideal forms fixed in the mind of God. Right understanding, in this reading, means looking at everything in nature as shadowy renditions of timeless, immutable originals. As we look at them, they will turn our minds and souls toward what is otherworldly, eternal, and therefore undiminished by the passage of time. Analogy, therefore, amounts to a metaphysics of the eternal present.

The analogical way of reading nature has influenced both theology and science. The American theologian Jonathan Edwards (1703–1758) offers an example of classical theology's analogical reading of nature:

> Why should not we suppose that [God] makes the inferior in imitation of the superior, the material of the spiritual, on purpose to have a resemblance and shadow of them? We see that

even in the material world, God makes one part of it strangely to agree with another, and why is it not reasonable to suppose He makes the whole as a shadow of the spiritual world? . . . If there be such an admirable analogy observed by the Creator in His works through the whole system of the natural world, so that one thing seems to be made in imitation of another, and especially the less perfect to be made in imitation of the more perfect . . . [why] is it not rational to suppose that the corporeal and visible world should be designedly made and constituted in analogy to the more spiritual, noble, and real world?[5]

The analogical stance has also infiltrated modern science. Science, understood analogically, is a matter not just of examining, counting, and measuring physical objects but also of looking for the timeless mathematical patterns that make things intelligible to human inquiry. The analogical way of reading nature lives on today especially in the speculations of contemporary theoretical physicists who consider the temporal world unreal in comparison with the timeless mathematical world, the realm of pure numbers in which the natural world imperfectly participates. Not a few mathematical physicists have enshrined timeless geometric forms as real while taking the passage of time to be subjective and illusory. There is, notably, a considerable measure of analogy in Einstein's love of eternity and in his never fully forsaken Spinozist way of understanding nature.

Einstein, like most modern scientists, used mathematics to make the universe intelligible. His use of non-Euclidean geometry opened up for him a new window onto the universe's previously hidden geometric coherence. His geometry, along with a refined imagination, led him to the surprising discovery, for example, that gravity and acceleration are somehow equivalent. Einstein wanted to know how things cohere, what brings unity to their parts. Intelligibility, unity, coherence—these are what to look for in any explanation. Einstein found what he was looking for through the medium of geometry.

Anticipation. Einstein's scientific delight in the unifying power of geometry led him to leave something vital out of his picture of the universe:

the passage of time. He tried not to ignore time, and he even factored a spatialized rendition of time into his geometric outline of the universe. But his subordination of time to geometry allowed him to ignore the fact that time flows and that it does so irreversibly. His obsession with geometry concealed from him the fact that the universe is still coming into being. His love of timelessness caused him to ignore the passage of time and the notion that the universe cannot yet be fully intelligible because it is not yet fully actual. It did not occur to him that the universe is a story of dramatic awakening for whose deepest meaning we can only wait.

Especially after Darwin, the world of thought started to awaken from its analogical slumber and its dreams of eternity. Because of revolutions in biology and contemporary cosmology, natural scientists now see nature not as an imitation of timeless forms but as a long passage in which older forms continually decay and new forms take their place without any sharp breaks. Types of living beings are born, flourish momentarily, and then vanish. Over a long span of time, different forms replace older ones in a continuous temporal transformation.

Natura non facit saltum. It is not nature's habit to make large leaps, but instead to unfold gradually. Because of nature's propensity for bringing about new beings step by step, therefore, we need an alternative way of reading the universe, one that differs from both archaeonomy and analogy. I call this alternative stance "anticipation." Anticipation claims that the meaning of everything in an unfinished universe lies not behind, in the past, or up above, in eternity, but up ahead, in the future, in the region of what is not-yet. Anticipation does not insist on full intelligibility here and now, because the story of the universe is not over. Intelligibility in our universe is mostly beyond our present understanding, because the universe is still aborning. To encounter its true meaning we have to wait—patiently, attentively, even intergenerationally. A good name for the disposition of patience is faith. And faith entails a metaphysics of the future: What is most real is not the past or the present but what is yet to come.

The way of anticipation, of course, needs also to be aware of what has happened in the past, that is, in cosmic history so far. Anticipation is not opposed to historical excavations by sciences that dig up the past. In fact, the anticipatory stance requires familiarity with the series of physical events that have led up to the outcomes now visible in the present. Anticipation follows eagerly the historical recovery of past states of nature by evolutionary biologists, geologists, astronomers, and astrophysicists. What it opposes is not archaeological excavation but archaeonomic explanation. Anticipation opposes the belief that nature, both present and future, can be rendered fully intelligible solely by meticulously tracing a series of efficient and material causes back into the cosmic past. Understanding, or making sense of things, means seeking their intelligibility. Things become intelligible to us, however, not simply by our laying out their past history or grasping how they have already been cobbled together but also by waiting to see how they will fit into future unifying syntheses. How could we ever have understood a carbon atom, for example, unless we had waited to see how, along with other atoms, it fits into a living cell in communion with other atoms?

Anticipation, however, does not assume that the universe is being drawn toward a predetermined goal. What I am calling anticipation is not the same as teleology or finalism, since it does not claim that the end of the cosmic process is already determined. Anticipation, as I am calling our third way of reading the universe, does not look for the implementation of a fixed divine plan. It reads the cosmic story instead as a long indeterminate awakening to something dawning on the horizon of the not-yet. Anticipation is thereby open to outcomes in the cosmic drama that may bring surprising illumination to the past and present.[6]

Anticipation, I propose, is the most realistic way of coming into contact with what is going on in a dramatic universe. The anticipatory stance requires a wayfaring patience, a virtue lacking in archaeonomy and analogy, both of which are eager to get to the finalized meaning of things forthwith. The archaeonomic and analogical readings of nature are unwilling to let the meaning of things arrive gradually, at its own pace and in its own time. Anticipation, by contrast, allows the universe to be an unfolding temporal drama for whose meaning we can only wait with active attentiveness. We read the cosmic story appropriately, therefore, only if, in each present moment, we wait with bridled expectancy. This means that we have to withhold any final judgments here and now regarding what exactly is going on in the universe. This is an instruction that neither analogical theology nor archaeonomic materialism is eager to hear.

The biblical figure of Abraham represents the disposition of anticipation and the virtue of heroic expectation.[7] Abrahamic anticipation means openness to the prospect that something mysterious and momentous may be taking shape up ahead as time goes on. This way of reading embodies both faith and hope. It seeks meaning, but it does not demand "to see the final scene" of the drama here and now. Patience—"one step enough for me"—is the way to arrive at a right understanding of a universe that is still coming into being.[8]

Anticipation does not look for the universe's intelligibility by going back to the past archaeonomically, nor by looking up above to heavenly archetypes purified of the passage of time, as analogy proposes. The anticipatory approach finds reasons for our hope in the precious outcomes we know as life and mind, already brought to us in the course of the history of the universe, but it does not dare to say exactly where things are going from here. The cosmos was not virtually finished at the start as archaeonomic materialism implies, nor is it a rough copy of a fully drafted plan in the mind of God, as analogical piety reads it. Anticipation reads the cosmos right now as a wide array of avenues for becoming *more*.

The Archaeonomic Stance and the Atrophy of Patience

Abrahamic patience is essential to an anticipatory reading of the cosmos, but the kind of patience that theology associates with Abraham has weakened over the centuries. In contemporary Western cultures, anticipation has been supplanted by analogy and archaeonomy, both of which are so impatient for present understanding that they have the effect of abolishing time. The resurgent atomism of modern scientific naturalism and the reactionary religious balm of otherworldly spirituality have jointly muffled the Abrahamic spirit of anticipation, with its deep appreciation of the transformative effects of duration. Both Democritean analysis and Platonic analogy, often in hybrid versions, live on in religions, popular culture, and intellectual life today. Archaeonomic atomism still hovers over the research agenda in universities all over the world, and it remains largely unchallenged. It remains today the prevalent academically and intellectually approved way of reading the universe. It has brought with it not only an impatient demand for immediate mathematical clarity but also a cosmic pessimism antithetical to anticipation and hope. I am not claiming that science itself is inherently impatient, since doing science rigorously requires enormous patience. It is archaeonomy, not science, that lacks this virtue.

Western theology, for its part, has for much of its history clung to the analogical way of reading the universe. The analogical stance has deeply influenced Christianity's ideas of God and fostered unrealistically perfectionist habits of thought and morality. It still remains out of touch with the protracted pace of an emerging universe. It has failed, I believe, to capture the anticipatory spirit of Abraham, the prophets, and Jesus. Theology's analogical reading of nature is unable to prepare religious hearts and minds to embrace the post-Einsteinian universe. When we look at nature, when we recite the Nicene Creed, or when we do theology in an exclusively analogical way, as is usually the case even today, we fail to appreciate the world's forward narrative momentum, as well as the human soul's need to have a new and surprising future. Analogy lacks the adventurous anticipatory accent of Abrahamic faith. So it cannot sense when something significant and unprecedented may be looming on the horizon of this world's temporal becoming. And it fails to satisfy the soul's need for a freedom that can thrive only in the open space of a universe that is somehow not-yet.

An exclusively analogical reading of nature often settles for a world-fleeing, time-suppressing, other-worldly optimism that ignores the human need for long journeys and far-off horizons. Analogy is satisfied with the shrunken expectation of liberation *from* the physical universe instead of hoping for the long liberation *of* the universe. Until recently, Christian theology, as a result of its close association with the analogical reading of nature, has cared very little about the final destiny of the physical universe or even about the long-term future of planet Earth. If, however, Christians were to interpret the Bible, the Creed, and the universe in the light of anticipation, everything would look different. First, they would understand that nature, contrary to the religious preferences of Spinoza and Einstein, can never again plausibly be identified with eternity and necessity. And, second, they would come to realize that the solution to our anxiety about perishing is not to flee from time but to align ourselves with the cosmic drama of an ongoing awakening.

Of our three readings, I believe that only anticipation lines up satisfactorily with the biblical roots of Christian faith, with the fact of irreversible time, and with the dramatic quality of the cosmos that science is bringing to our attention. The universe that keeps flowing on beneath the abstract quietude of Einstein's geometrical formalism, is gradually awakening. Any meanings the universe may now be carrying cannot be captured by archaeonomy or analogy. Neither of these perspectives has the patience to look for what may be blossoming narratively up ahead. Indeed, neither the archaeonomic nor the analogical stance is prepared even to notice the awakening that has been going on in the universe for billions of years.

If we take an anticipatory stance, however, whenever we look at the series of stages the cosmic journey has gone through up until now, how can we help noticing that, at every past moment in the long cosmic passage, something new and momentous was silently and invisibly beginning to form up ahead? We have no good reason, therefore, to assume that the faithful arrival of new forms of meaning in the future is any less likely today than before. Maybe, after all, there are very good reasons to hope.

How so? If any of us had been present shortly after the Big Bang, during the period in which the universe was a monotonous soup of preatomic elements, neither archaeonomy nor analogy would have prepared our minds to encounter what was going to happen later on. Only if we had adopted the anticipatory stance of waiting attentively would our thoughts have made a place for what was coming. Had we been present during the earliest period of cosmic time, when the universe was in a state of elemental diffusion, only an anticipatory disposition would have shaped our inquiries to expect the later epochs of life, mind, and freedom that were then hidden in the future.

So we need not despair right now, at this present moment of cosmic and human history, if we cannot yet make complete sense of our universe.

It is enough to know that something like an awakening has been happening in the cosmic story up until now. Perhaps the activity going on in our minds at the present moment, as we seek meaning and truth, is indicative of a cosmic awakening still in progress. We should have learned by now, simply by pondering past history, that an awakening universe requires, on our part, a willingness to wait, sometimes for long periods of time, to tell whether something big is beginning to emerge. If so, anticipation, not archaeonomy or analogy, is the true realism.

Archaeonomy is content to understand things only in terms of what came before, and analogy views present phenomena as imperfect copies of a perfection eternally established up above. Anticipation alone looks steadily toward what is still coming. It is in the spirit of anticipation that Christians, I suggest, may also now recite the Nicene Creed after Einstein. Anticipatory faith expects to be drawn forward—along with the rest of the universe—into the ever-replenishing mystery of the not-yet. It is anticipation rather than archaeonomy or analogy that makes room for a creator "who makes all things new."[9] As we shall see later, the emergence of human freedom allows room also for the tragedy of our refusing to participate in the further awakening of the universe and even for our nihilistic will to destroy life-systems billions of years in the making.

Rather than professing our faith solely in the spirit of the analogical stance that so heavily influenced early church councils and subsequent theology, we can, with a cosmologically informed faith, allow our thoughts, aspirations, and actions to be shaped by a sense of nature's promise. An anticipatory faith does not instruct souls to separate themselves from matter and time; instead, it generates hope for a future that includes the destiny of the entire universe. A sense of future cosmic transformation consistent with hope was not part of Einstein's personal religion. His spiritual attachment to eternity was shaped more by analogy than by anticipation. Nevertheless, anticipation seems consonant with the narratively shaped cosmos given to us by Einstein's science. In the Nicene Creed "we look forward to . . . the life of the world to come," but after Einstein, faith looks not for the abolition of time but for its dramatic fulfillment. Accordingly, people of faith may look for God, not in interruptions of nature, but in a narrative coherence that is still incomplete. The "invisible" world that God creates consists not only of what is hidden spatially at present but also of what belongs to the anticipated domain of the not-yet.

Summary

Science after Einstein now allows us, I believe, to recover the long-lost anticipatory matrix of early Christian faith. Beneath the mathematical abstractions of Einstein's modeling of mass, space, and time we find a universe revealing itself as a long and far-from-finished story. Like any story, the universe may have meanings that are just barely beginning to take shape. Cosmic meaning, in that case, remains presently out of our grasp, since the story is not over. In the context of our thirty-volume cosmic narrative, therefore, the function of faith and Creed is to turn our attention forward, toward the horizon of possible future volumes yet to be written.

Understanding nature narratively—as a drama of awakening to indestructible rightness—allows for a new kind of conversation between science and Christian theology. Today, the main objective in this exchange is not to specify physically whether or how divine action does or does not interrupt nature's regularities. Rather, the objective is to learn how to read the ongoing story of the universe. Currently there are three main ways of doing so. These are archaeonomy (a contemporary version of scientific materialism), analogy (the approach of classical theology), and anticipation (the approach I recommend for theology after Einstein). Our three ways of reading natural history correspond to three ancient perspectives on the world, those, respectively, of Democritus, Plato, and Abraham.[10]

SIX

Origins

> The man who is thoroughly convinced of the universal operation
> of the law of causation cannot for a moment entertain the idea
> of a being who interferes in the course of events.
> —Albert Einstein

> The grandeur of the river is revealed not at its source but at its estuary.
> —Pierre Teilhard de Chardin

> When, in the beginning, God created . . .
> —Genesis 1:1

UNTIL NOW MOST THEOLOGICAL reflection on Einstein's science has focused on the Big Bang and what it implies for the doctrine of creation. The new cosmology ushered in by the science of relativity seems, at least on the surface, to support the Christian belief that the universe had a beginning in time. And is it not the case that whatever begins to exist in time, can do so only if something that already exists brings it into existence? So isn't the idea of a divine creator a reasonable explanation of why the universe—which seems not to have existed eternally—had a beginning?

No doubt someone will ask, How, then, did God the creator come into existence? That question, however, makes no sense, because we are trying to explain whatever begins to exist, not what has no beginning. God, according to Christian theology, is the uncreated creator who always exists and so could not have begun to exist. What is at issue here is not whether God had a beginning but whether the universe had a beginning.[1]

Did the universe really have a beginning? A lot seems to ride on this question. Cosmologists today agree that the universe is expanding, and this points to its having had a beginning. For if we follow the cosmic story back in time, the lines of its expansion get closer and closer together the farther back we go. Eventually, if we go all the way back in time, we arrive at an infinitesimal point (13.8 billion years ago) at which the cosmos seemingly began to exist, perhaps as a fluctuation in a quantum vacuum.

Scientists still do not know exactly what went on during the initial microsecond of the Big Bang universe's existence. The known laws of science do not apply during that moment. Furthermore, the existence of an original quantum vacuum—if that is what brought the universe into existence—needs an explanation too. In the biblical book of Genesis, God is said to have created the world "in the beginning," and a large number of scientists and theologians consider Big Bang cosmology at least consonant with, if not proof of, the existence of an eternal creator, God.

Christian theology, however, does not have to worry about whether the universe had a beginning in time, because the doctrine of creation is not about temporal beginnings. It is about the dependency of the universe on a gracious source of being other than itself. If questions about origins are pursued in depth, they eventually lead back to the theological question of why anything exists at all. That is the real question.

Beginning before Einstein

Even independently of Einstein's theory of relativity, astronomical considerations have raised doubts about whether the universe has always existed. In the first place, if the universe has been around forever, would it not have collapsed upon itself by now? For if every physical body tugs gravitationally on every other, as Newtonian physics stipulates, and if the amount of time for tugging is limitless, then the universe would by now have become one big lump of locked-in mass. Instead, great distances still separate galaxies, planets, and stars. So isn't the universe finite in age? And if so, didn't it have a beginning? Is it not reasonable to conclude that the cosmos has been brought into being by something other than itself—perhaps by the creator God of the Bible?

The universe is at least billions of years old. Isn't that enough time for gravity to have pulled all the heavenly bodies into a single lump? What keeps the planets, stars, and galaxies apart? This puzzle has yet to be solved to everyone's satisfaction. Einstein, at least for a time, thought the physical universe was eternal, and he speculated that it must have always had a cosmological constant, an antigravity factor, to keep the heavenly bodies apart, thus preventing the universe from collapsing onto itself.[2] If so, the universe does not need to have had a beginning.

Einstein's own physics, however, as interpreted by Georges Lemaître and others, appears to have made the cosmological constant unnecessary. The widespread distribution of heavenly bodies is caused by the ongoing expansion of the universe. The universe is old but still finite in age. It is still expanding, and the speed of its spatial expansion overpowers the pull of gravity. If the universe had a beginning in time, Einstein's cosmological constant, therefore, is not needed to explain the great distances separating stars and galaxies. The cosmic expansion rate is enough. This seems compatible with the existence of a God who created the universe a finite amount of time ago.

Perhaps, though, Einstein's notion of a cosmological constant is not totally wide of the mark. Presently, with the benefit of increasingly precise measurements of stellar distances and motion, the universe seems to be not only expanding but doing so faster and more unevenly in its pace of acceleration than scientists had previously calculated. The additional speed of expansion functions something like Einstein's cosmological constant. Astrophysicists now attribute the unexpected boost in acceleration to dark energy, a still-obscure physical factor that changes the rate of cosmic expansion beyond previous expectations. Some astronomers have recently suggested that the extra acceleration implies that the universe may be a billion or so years younger than the standard figure of 13.8 billion years.[3] Nonetheless, this modification does not in principle rule out the possibility that the universe was created in time and that it had a definite beginning.

A second consideration should also have made scientists before Einstein question the assumption that the universe has no beginning. If the universe had existed everlastingly, why is our own planet not flooded with light from the innumerable stars distributed throughout the vastness of space and time? This conundrum is known as Olbers' paradox. What is paradoxical is that the earth gets dark at night. Light travels at a high but finite rate of speed. If the universe had existed forever, for all eternity, the light from billions of stars would have had plenty of time to flood our planet by now; no star would still be out of sight because all visible light would have already reached us. In other words, nighttime here would not exist. But our sky gets dark at night, which must mean that the universe is not eternal, that it had a beginning, and that it has not been around long enough for our planet to be lit up all the time. Doesn't the biblical idea of a God who created the universe at the beginning of time provide a reasonable explanation of why the age of the universe is finite and why nighttime on Earth exists?

Third, if the cosmos had existed forever, it would have run out of energy by now. Physics tells us that the overall amount of usable energy in the universe is constantly decreasing. If the universe had no beginning and if it were everlasting in duration, it would have run out of usable energy long ago. The loss of available energy, or entropy, is always increasing, which means that an infinitely old universe would be energetically dead by now. Why isn't it? Maybe because it has not been around forever? The point is that, even independently of Einstein, scientists could have reasonably assumed that the cosmos is limited in age. If so, is there a more reasonable explanation for why the universe began to exist than the idea of a generous creator God?

My theological instincts tell me not to be too hasty in drawing this conclusion. First, the plausibility of divine creation does not require that the universe had a beginning in time. Second, theology needs to allow purely natural explanations to be taken as far as they can possibly go. It is a scandal that theologians and even some scientists have sometimes brought in theological explanations as a substitute for purely scientific accounts of natural phenomena. And, third, theology needs to focus primarily not on the question of cosmic beginnings but on the fact of entropy and the possible end of the universe.

Entropy implies that the universe will eventually run out of available energy, thus bringing an end to life, consciousness, and cultures everywhere. Isn't entropy, then, a reason for cosmic pessimism? Entropy seems like bad news, since it implies that the universe will eventually run out of usable energy and die. So the question of why and how the universe will end is of no less interest to lovers of time—and that must include Christians—than the question of why and how it all began.

Sooner or later, as most scientists now agree, the universe will terminate in thermodynamic equilibrium, and the story will be over. But there is another way of looking at entropy and its implications. As I mentioned earlier, entropy is the main physical reason why the passage of time is irreversible and why the universe has the shape of a story. It is the unidirectional movement of time that allows the physical universe to be a great drama of awakening. Life, thought, and culture would not have appeared in the universe apart from the one-way narrative freeway laid down by the increase of entropy. Entropy implies that the cosmic journey is headed toward death ages from now, but does this mean that nothing can be harvested permanently from the drama of cosmic transformation now going on? Isn't it possible that the indestructible rightness to which the universe is awakening can take into itself and save the content of the story it has set in motion? Or will nothing, in the end, remain of the cosmic drama of awakening?

In the history of human thought, people have often dealt with the threat of final perishing by seeking a return to origins. We humans are almost archetypically attracted to origins because "in the beginning" it seems that an exceptional pulse of power made the world exist by overcoming the threat of nonbeing.[4] Since existence in time means that we and the universe are always subject to the threat of nonbeing, it is not surprising that we would want to reconnect with that primal power by going back to the beginning. Where else, we might ask, can we find the potency to expel perishability from nature and our lives except in the beginning—before cosmic time began?

Much of religion consists of a nostalgia for the "strong time" of origins—the religion scholar Mircea Eliade's term.[5] For thousands of years we humans, especially by way of our religions, have sought out a "ground" apart from the diminishment of time and tried to replant our perishable lives as well as the universe itself in that sacred soil.[6] In our myths and religions we have demonstrated how haunted we humans are with the theme of origins. Religions provide symbols, stories, and elaborate rites that allow us to tap into the indestructibility from which the perishing world is thought to have strayed. The mere phrase "in the beginning"—*in principio, en arche*—can bolster our courage, especially if it is accompanied by scriptural and ritualistic maps showing us how we may return to that strong "time before time."[7]

A major objective of theology after Einstein, then, is to examine whether such maps are reliable. Following the typology outlined in the preceding chapter, we may start our survey by looking comparatively at three ways of thinking about cosmic origins in terms of the new scientific cosmic story.

Archaeonomy. The archaeonomic stance taken by most contemporary scientists and philosophers today is a good example of a longing, mostly unconscious, to return to origins. To the strict archaeonomist the universe was already loaded in the beginning with everything that was ever going to happen later on. Archaeonomists cannot help being deeply enchanted with the quest for cosmic origins, because their assumption is that the fullness of being always resides "back there." In the original physical state of the cosmos, archaeonomists assume, the seeds of whatever was going to happen later on had already been sown. The universe needed only the long passage of time to lay out its primordial treasures step by step. Returning to origins, for the archaeonomist, means tracing the passage of time back to the beginning. Since there is no possibility of our literally going back that far in time, we may presently connect with cosmic origins by breaking present organisms and other objects down into their most elemental parts. By way of this analytical reduction, archaeonomists hope to approximate what the universe looked like in the beginning.

Like our mythmaking ancestors, archaeonomists are secretly in search of the strong time of beginnings. Archaeonomy, by figuratively pitching its tent at the beginning of cosmic history, is an almost religious way of keeping at bay the threat of nonbeing that now threatens everything in time. By locating the fullness of being at the point of origins—just before the dangerous march of moments is taking off—archaeonomy clings unknowingly to the angel of timelessness.

True, the passage of time spreads out the world's original content over an immensity of space and time, but to the archaeonomist this distribution is metaphysically incidental. Everything that happens in the universe is merely the uncoiling of a fullness already latent at the outset. Even though eventually the original mindless elements get transformed by impersonal laws into anxious and curious human minds billions of years after the beginning, nothing truly significant will have happened, or will happen, that was not already part of the original package. The universe was lifeless and mindless *en arche,* and it is back to that original state of pure elemental simplicity and deep sleep that time will return in the end. In the final analysis, nothing intrinsically important will ever have been accomplished. The universe will have been a mere series of states, not a drama of awakening.

To the archaeonomists, nothing really new, but only the appearance of novelty, can be added to the original state of being. The store of usable energy present in the beginning will eventually run out, and then the whole universe will be thrown back into timeless slumber. The slope of entropy will have been flattened completely, leaving no place for "before" and "after." More important, no room will remain for what is not-yet. Anything that happens after the universe's inauspicious beginning can be nothing more than the working out algorithmically of the mindlessness that came first.[8] Instead of waiting for unprecedented cosmic meaning to become manifested in the future, archaeonomy keeps turning back to the bare beginning. Archaeonomists admit that they could not have described exactly what was going to happen after the beginning, but they are sure that whatever has happened was predestined to do so at the beginning of time.

The salient feature of archaeonomy, therefore, is its implicit denial that the cosmos is a story and that time has any lasting significance. Nothing is really working itself out in the universe. The passage of time adds nothing significant to the completeness of what was already there in the beginning. What kind of comfort, then, can archaeonomy offer to those tender souls beset with the anxiety of final perishing? The British poet Algernon Charles Swinburne gives one response:

> From too much love of living,
>   From hope and fear set free,
> We thank with brief thanksgiving
>   Whatever gods may be
> That no life lives forever;
> That dead men rise up never;
> That even the weariest river
>   Winds somewhere safe to sea.
>
> Then star nor sun shall waken,
>   Nor any change of light:
> Nor sound of water shaken,
>   Nor any sound or sight:
> Nor wintry leaves nor vernal,
> Nor days nor things diurnal;
> Only the sleep eternal
>   In an eternal night.[9]

Analogy. The naked fact that nothing in time lasts makes philosophers weep, causes religions to separate God from nature, persuades sensitive thinkers like Spinoza to embrace pantheism, and leads scientists like Einstein to treat the passage of time as an illusion. If time never really flows, then maybe perishing never really happens either. Einstein's sensitivity to transience and loss, in my opinion, led him to hold on to the idea of an eternal universe as long as he could. Since time implies decay, its swift passage leads us to look for ways to escape it. Einstein's way of doing so was to factor temporal duration out of existence by way of geometry.

To analogical theology, however, the human concern for origins cannot be satisfied by finding a first moment in time. The real question is why there is anything at all rather than nothing. This is a metaphysical issue, not a scientific one. It is the universe's existence, not its temporal beginning, that needs explaining, and science is out of its depth in trying to do so. Some scientists plunge into metaphysics anyway, confusing it with physical science. The Cal Tech physicist Sean Carroll, for example, declares that "if and when cosmologists develop a successful scientific understanding of the origin of the universe, we will be left with a picture in which there is no place for God to act."[10] Followers of analogy would respond that in taking up the question of God, physics has abandoned its former methodological purity. The conflation of metaphysics or theology with physics has never been good for either discipline.

Some scientific naturalists try to avoid theology by multiplying universes *ad infinitum* in order to increase the probability that our own life-and-mind-bearing cosmos could pop up naturally, purely by chance rather than by divine design.[11] Analogical theology, however, replies that this effort merely postpones facing the real issue. The existence of a multiverse is no less in need of ultimate explanation than the Big Bang universe is. So the Nicene Creed's claim that God is the "maker ... of all things, visible and invisible," is still satisfying to those who take the question of origins metaphysically rather than physically. To analogical metaphysicians the purely naturalistic pursuit of cosmic origins is perhaps an interesting diversion but not an intellectually serious response to the question of why anything exists at all.

Analogy is both spiritually and intellectually seductive. The constant threat of nonbeing has persuaded Christian thinkers to reserve a place for both souls and God in a sphere of timelessness apart from the material world. Analogical theology has been so widely appealing because in effect, it promises a final abolition of time. Even the recent scientific dating of cosmic origins at 13.8 billion years ago makes little difference to the analogically minded. It does not matter how long or short the passage of cosmic time is, provided that souls at some point can find a way to escape temporal duration altogether and move on to eternity.

Analogical spirituality assumes that the point of our lives—and of faith—is to let ourselves be freely captured by eternity so that we may contemplate forever the infinite beauty of God. Terrestrial time exists mostly as a school for soul-making. Given the fact of perishing and given the shortness of our lives, the analogical vision, with its promise of vertical liberation from the temptations and terrors of time, still seems irresistible to millions of religious believers. It is not surprising that soon after the death of Jesus the analogical philosophy already circulating in the Mediterranean region became the standard intellectual backdrop for Christian spirituality, where the latter has lived for centuries.

Early Eastern Christian writers, such as Clement (150–215), Origen (184–253), Athanasius (296–373), Chrysostom (347–407), Basil (330–379), and Gregory of Nyssa (335–395), affirmed human destiny as participation in the bodily resurrection of Jesus, but their Platonic leanings also led them to spiritualize Christian hope as an awakening to the state of pure timelessness. In the West, the influence of Plato also helped shape the otherworldly hopes of Augustine (354–430), Anselm (1033–1109), Aquinas (1225–1274), and their followers. Analogical theology, as I see it, still serves as the conceptual backbone of most Christian spiritual life, and I have no doubt that it will comfort souls for generations to come. In its diminution of cosmic time, however, analogy dampens Abrahamic hope and strays from Christian faith's emphasis on God's incarnation in time. The more tightly Christian theologians cling to prescientific analogical metaphysics, the less likely they are these days to look for any theological meaning in deep cosmic time. They are less concerned about the dramatic awakening *of* the universe than about the soul's awakening *from* the universe.

Given how hard life is for most of us, it is understandable that we sometimes long to exchange time for eternity. To shield ourselves from the prospect of living fully in time we may be ready and willing, even in the best of circumstances, to leave the physical universe behind for good. Analogy, for that reason, continues to have its appeal. Nevertheless, it lacks the conceptual and spiritual breadth to give us a hope proportionate to the enormity of time's long passage. It lacks the patience to give the universe time enough to become fully what it has the potential to be. It finds meaning in nature's fleeting unveilings of the eternal present, but it has no great love for irreversible time or the idea that the universe is still being created. Analogy does not formally dismiss or inevitably despise the fact of time, but it has little interest in looking into the cosmic story for dramatic suspense or narrative coherence—or for new opportunities for human creativity provided by time's vast duration. The long, winding journey of the cosmos in time means little or nothing to those whose attention is fixed mostly on what lies either in the beginning or in eternity.

Teilhard de Chardin, the renowned geologist and Jesuit priest, was one of the very first scientists in the twentieth century to realize that the universe is a story and that nature is still aborning. Theology, he rightly insisted, should not avoid the question of what cosmogenesis, the growth of the universe, means for our sense of God. He was puzzled that his fellow Christians habitually preferred a spirituality rooted in analogy over one rooted in anticipation. In 1916, while Einstein was issuing his ideas on general relativity and reducing time to geometry, Teilhard was complaining that the analogical vision was still leading his fellow Catholics to "rise up to Heaven, borne on the wings of contempt of the world." In trying to escape from time Christians had forgotten that they "still needed the nourishment of earthly bread." And so, when they finally realized that they could not live on otherworldly hopes alone, they discovered that "their hands had lost their skill and their hearts had lost their enthusiasm."[12]

Anticipation. Anticipation, unlike either archaeonomy or analogy, is attuned first and foremost not to the cosmic past nor the eternal present but to the refreshing mystery of the not-yet. It cannot separate the quest for cosmic origins from the more important concern for the cosmic future. For that reason, anticipation patiently follows the cosmic story, not to determine how the world began or when and how we may escape from it, but in search of a meaning that has yet to become fully actualized. The anticipatory approach is fully aware of science's informative excavations of biological, geological, and cosmic history. But anticipation's fundamental observation is that the universe has yet to become fully real or fully intelligible. Only an anticipatory stance, not the quest for origins, can let us meet up with the essence of the universe in its long temporal awakening.

Disciples of archaeonomy and analogy still dominate the world of thought, and even though they are opposed to each other in most respects, they share a sense that time's immensity has little value or meaning. Archaeonomic atomists follow the course of cosmic history backward in time to an original spread of elements and physical laws. They claim that whatever was going to happen after such a spare beginning was already locked into the original synthesis. Deep time, for the archaeonomists, is the pointless unfolding of what was compressed into the initial packet. Everything that happens in time is exactly what we should have expected all along. Time has no meaning, then, other than gradually letting us in on what had already been virtually finalized.

In current attempts to make sense of deep time the voice of anticipation remains barely audible, even though its appreciative sense of irreversible time was shaping hearts and minds for centuries prior to the birth of Christianity. Seeking an escape from time was foreign to the anticipatory faith of Abraham, Moses, the prophets, and Jesus. What their faith sought was not an escape from time but time's fulfillment.

Time, at least in prophetic faith, is not to be avoided but cherished as the necessary medium for the swelling of expectation. Time is something to be indwelled, not sloughed off; and deep time is a fertile field for nurturing an ever-more-expansive anticipatory faith. The mystery of God, in this setting, is not to be explored primarily by returning to origins. Nor does the mystery of God need to be located up above, or hidden inside human subjectivity. These may be important portals to the experience of God, but as I have been proposing, we may look for God especially in the unfathomed depths of the hidden future to which the universe is now awakening. It is to the Absolute Future that we may still lift up our hearts.

Archaeonomy and analogy are both tailored to fit our unwillingness to wait. They expect intelligibility to be immediately available, whether by sight, flight, geometry, or logic. For archaeonomy, nature has never seemed alive with expectancy, since the fullness of being was already bottled up in the first moment of time. Everything else in cosmic history is, then, merely a spinoff of an original pointlessness. Analogical theology does not prefer to wait, either. In our state of exile, analogy implies, the flush of springtime and other glories of nature may remind us of timeless Eden before the Fall, and they may point our hearts toward an eternally unchanging deity.[13] Analogy briefly lets in the light of infinite divine beauty. It fails to notice, however, that the glories of nature are also promissory hints of dramatic transformations yet to come.

Einstein passed over the wondrous mystery of the not-yet because of his two-timing embrace of archaeonomy and analogy. Focusing only on geometry, he never bothered to ask why the universe is a long-drawn-out drama of awakening. His Spinozist philosophical bias led him to favor the idea of an essentially immutable universe that pinches the passage of time out of existence. The emphasis he gave to his own timeless geometry of spacetime distracted him from the universe's actual narrative openness to the future. Theology after Einstein, however, no longer needs to take religious flights into eternity as the solution to the perpetual problem of transience. If the healing embrace of God receives all of time into the divine life, there is no longer any need for us to vanquish time or to flee from it. Instead, we may look into the whole story of nature for indications of the promise of a new future up ahead for all of creation.

Consequently, it is not primarily because the universe may have had a beginning in time that God is an acceptable topic for theological discussion after Einstein. Rather, it is because the universe is always being given a new future. According to Christian faith, an indestructible self-giving love is the creative source of the universe. But love does not force. Patiently it takes its time. It is not by dictating, but by attracting and letting be, that divine love allows creation to happen. That the ultimate source of cosmic existence and coherence is somehow not-yet is not a sign of divine weakness or deistic abdication. It is the trace of a unique kind of power—the power to let something other than itself come into existence and then to become even more.

Indestructible rightness itself—without the slightest loss to its superabundant generosity—undergoes an inner transformation as it gathers everything that happens in time into itself everlastingly. Almighty-ness, in this sense, allows for the flow of time and thus for the existence, flourishing, and freedom of creation. To address the question of how things came to be in the beginning and how they hang together intelligibly now, we need to keep turning toward the future. In the cosmic future we anticipate the arrival of an intelligibility not fully present in the beginning or now. As Teilhard wrote, "The grandeur of the river is revealed not at its source but at its estuary."

That the universe is comprehensible—that it fits the contours of our minds, as Einstein rightly emphasized—should fill everyone, especially scientists, with a sense of wonder. But why the universe exists at all and why it has a temporal/dramatic constitution are two questions that did not seem of great interest to Einstein. Even though he was wonderstruck that the universe is intelligible, he never seems to have felt fully the "ontological shock" that anything exists at all. And his preference for an unchanging universe prevented him from ever feeling the "narrative shock" of belonging to a universe that is still coming into being.[14]

Summary

Humans have always wanted to know how things came to be. How did life originate? Why are there so many species of life? How did evil and suffering get into the world? How did humans come to exist? How did consciousness enter into the picture? How did sexual behavior begin? And, above all, where did the cosmos come from?

These are all legitimate questions, but absolute beginnings are beyond our intellectual reach. This, I believe, is because the world is shaped not by design but by promise. At the time of its origins 13.8 billion years ago the Big Bang universe was already gathering itself into the narrow bottleneck of physical conditions and constants that would render the advent of life, mind, and freedom probable if only time would turn out to be deep enough. Physical transformations during the cosmic story—such as the primordial formation of free hydrogen atoms and the later, though still billions of years ago, cooking up of carbon in the hearts of stars—are essential stages in the drama of cosmic awakening. The gravitational constant and the cosmic expansion rate at the beginning had to fall within exceedingly narrow mathematical ranges if life, mind, freedom, faith, hope, and compassion were ever to come into the cosmic story.

If, moreover, there happen to be other universes beyond our own, worlds inaccessible to our present powers of observation, they, too, are at least remotely complicit in the dramatic awakening of our own universe. Of course, they would also have a metaphysical continuity with our Big Bang universe simply by sharing in the wondrous act of existing. And theologically, no matter how many universes there are, they are all part of the one creation, and they may also have a narrative togetherness in being statistically necessary for the existence of our own life-bearing universe. Perhaps, then, what is remarkable is not so much that a multiverse makes the existence of our own life-bearing universe mathematically and statistically intelligible, as archaeonomy holds. Rather, from an anticipatory perspective, it may be that the existence of our own awakening universe brings a dramatic intelligibility to the existence of a multiverse.

An anticipatory theology embraces deep time as a gift filled with meaning. As the universe awakens to the invitation of God, divine patience awaits the development of the universe. Time and the story of the universe are internal, not external to the divine life. So, whatever happens in time makes a difference to the very identity of the God who waits. A central Christian intuition is that time matters to God. Otherwise, it is hard to imagine that God truly loves and cares for creation. Analogical theology, to the extent that it cleanses the creator of contamination by time, has failed, in my opinion, to give us a God who fully loves the world. Analogy fails to acknowledge either the dramatic awakening of the universe or the intimate relationship of that drama to the inner life of God.

Anticipation looks for the meaning of creation not at the beginning of time but in the not-yet of what is coming. Looking for the date and description of an assumed point of cosmic origins long ago is both unnecessary religiously and futile intellectually. There is no pure point of origin or pure past to be found in a universe grounded in the not-yet. It is not past physical causation but the power of the future that calls the world into being. All attempts to pin down a discrete point of cosmic origins in the distant past are destined to frustration because there never was a moment in time when the universe was not already in the grasp of the elusive power of the not-yet.

SEVEN

Life

> It is the essence of life that it exists for its own
> sake, as the intrinsic reaping of value.
> —Alfred North Whitehead

> I believe in the Holy Spirit, the Lord, the giver of life.
> —Nicene Creed

WE HUMANS HAVE ALWAYS found ourselves suspended in time and space somewhere between the small and the large. If we measure things on a spatial scale stretching from the infinitesimal to the immense, we are average. Yet if we measure things in terms of their physical complexity, we are not average at all. On a curve ranging from the least complex to the most complex things in the universe, we are close to the summit.[1] Our organic and cerebral complexity, moreover, has recently introduced into the universe the marvel of thought, causing an immeasurable new jolt in the drama of cosmic awakening.

The immensity of space and time has been matched by the numerical quantity of atoms, molecules, cells, and the even more plentiful interconnections among particles, stars, plants, animals, humans, and ecosystems. And now, as a result of developments in human culture and technology, a new epoch of rising complexity is emerging over our heads and behind our backs: by way of scientific invention and human creativity, our planet is now weaving around itself a digital complexity whose growth appears to have no upper limit.[2]

This phenomenal rise in complexity is another indication of the irreversibility of time and the narrative shape of the universe. As the physicist Lee Smolin explains:

Complexity is improbable. It requires explanation. Nothing can jump immediately from a simple to a very complex organization. Great complexity requires a series of small steps. These occur in a sequence, which implies a strong ordering of events in time. All scientific explanations of complexity require a history, during which the levels of complexity rachet up slowly and incrementally.[3]

Here, though, I am also thinking of complexity as a third domain of cosmic immensity alongside those of space and time. A smaller, younger, and more numerically undersized universe could not have sponsored the gradual rise of physical complexity required for the existence of life. The arrival of life in the universe allows us to see now how mutually interwoven the three immensities — time, space, and complexity — are. The emergence of physical complexity in cosmic history could not have occurred if space, time, and absolute numbers were not immense. Only within an almost limitless spatial, temporal, and numerical expanse can sufficient opportunities be made available for increasingly complex arrangements of elements and interconnections to arrive step by step in time. It is on the wings of our universe's irreversible temporal journey toward increasing physical complexity that life arrives, survives, and develops.

The arrival of life around 3.7 billion years ago has given dramatic intelligibility to the three cosmic immensities. If we had been following the cosmic story from its deepest past until now, we would have had to wait until life arrived before we could make good sense of the cosmic immensities established earlier. Until recently, however, most scientists interpreted the seemingly wasteful expanses of space and time as proof of the universe's fundamental hostility or indifference to the emergence of life. Spatial and temporal immensity appeared to have only a weak and extrinsic relationship to the existence of life.

Archaeonomy, which by definition is blind to the narrativity of nature, has looked at the emergence of life as a disturbance of the more natural lifelessness of nature. Not too long ago the molecular biologist Jacques

Monod, for example, claimed that life does not really belong to the spatiotemporal universe, which means that we humans are not supposed to be here either. Like many other modern scientific thinkers, he made the universe's apparent indifference to life the starting point of a tragic philosophy of nature. From Monod's archaeonomic point of view, life looks like an unintelligible anomaly in a vast cosmic emptiness.[4]

Recently, however, astrophysicists and cosmologists have begun to acknowledge that the universe's spatial, temporal, and numerical immensity is tied essentially, and not accidentally, to the drama of life. The universe is not an indifferent void in which a brief spark of life flashes momentarily only to be soon snuffed out. Instead, as we now look at nature in narrative retrospect, the three cosmic immensities are jointly predisposed to hosting the story of life. The universe is not hostile to life after all. The existence of life now renders the silent cosmic immensities intelligible in a narrative way that the pure archaeonomist is not prepared to notice.

Of course, we could never have predicted the existence of life simply by studying the three immensities in fine detail beforehand. Yet once we realize that life is an integral part of an unfinished story, the three cosmic immensities become intelligible: their coexistence makes sense as a condition essential to the configuring of a dramatic universe. We not only appreciate today the special geometric coherence that Einstein brings to our understanding of spatial and temporal immensity but also the dramatic coherence that makes cosmic immensity an essential part of the story of an awakening universe.[5]

What Makes Life Dramatic

The arrival of life, I have said, brings dramatic intelligibility to the universe. Life is dramatic, as the scientist and philosopher Michael Polanyi has observed, simply because its defining attribute is that of *striving*. Striving, seeking, straining, trying, groping, aiming, resisting, struggling to accomplish something—these are distinguishing marks of life.[6] Life involves much more than striving, I realize, but it is especially their anticipatory quality that lets organisms stand out clearly as alive instead of lifeless. In life's most primitive forms striving may be negligible, but as life grows in complexity this quality becomes increasingly more manifest.

From both the archaeonomic and the anticipatory perspectives the sharp lines that people usually draw between matter and life seem blurry, but for a different reason in each case. The archaeonomic perspective pictures nonlife as dominant, as eventually resolving the whole cosmos into a cloud of dust. The anticipatory perspective I am following in this book, on the other hand, blurs the line by dramatically linking the earliest and most elemental stirrings of matter to the striving of life.

We each first encounter the fact of striving in our own personal aims and struggles, but the awareness of our own striving allows us to intuit a special kinship with all other living beings. They are fellow strivers, and we share with them the anticipatory quality of a cosmos that joins all of life tightly to a drama of gradual awakening.

Along with the capacity to strive comes the prospect of either succeeding or failing. Unlike purely physical or chemical processes, living beings aim to succeed in one way or another, but their aiming for success is not always rewarded. As often as not, living beings fail to achieve their goals. Ultimately, all living beings fail to survive, and the prospect of this final failure makes life especially dramatic. The prospect of failure brings an element of pathos into the universe and introduces a dramatic motif into the cosmic story that becomes increasingly explicit as life becomes more complex.

Life requires that organisms keep on striving if they are to stay alive and evade death. Slacken the striving, and life ebbs away, as we know from our own experience. Long before the universe gave rise to human minds capable of groping for meaning and truth, it had already become explicitly dramatic. In the story of life, even in the earliest stirrings of some cells and organisms, the cosmos had become conative (striving) long before it became cognitive. Cognition itself is a highly advanced kind of striving. We humans spend much of our lives striving for understanding and truth (as you are doing right now). From the point of view of physics and chemistry, the arrival of life did not produce a ripple in the regularities of nature;

rather, a whole new kind of existence came into the universe, one in which beings can strive and achieve—but also fail.[7]

With the coming of life in the cosmic story, anticipation burst out into the open. Humans are perhaps the chief instance of life's anticipatory striving, but by no means are they the only ones. Even an amoeba anticipates finding nourishment. No celebratory parades or blasting trumpets announced the entry of anticipation into the cosmos. Nonetheless, the dramatic capacity for anticipation was being silently spun out of the three immensities of the universe starting long before life actually arrived. Physically speaking, the arrival of striving beings in our corner of the universe 3.7 billion years ago was so unobtrusive that the sciences of astronomy, physics, and chemistry would scarcely have noticed. Dramatically speaking, however, when life first came onto the scene, the whole universe had suddenly embarked on a remarkable new phase in its long awakening.

When living cells first made their appearance on Earth, it was not as though the cosmos was completely unprepared for their arrival. For even if life's debut was physically unimpressive when observed from a large-scale cosmological point of view, and even if life started off at a remote location on one lonely planet, its arrival gave our threefold cosmic immensity a dramatic meaning, in retrospect, that could not have been captured by the geometry of general relativity. Moreover, explosive sidereal events that transpired in space and time billions of years ago, as astrophysics has now shown, were not incidental to the fashioning of living complexity, anticipatory minds, and morally striving beings here on Earth and perhaps elsewhere. Physical and chemical transactions that went on in massive stars long ago were not extrinsic stagings but happenings internal to the drama of life. Astrophysical paroxysms that occurred in cosmic antiquity still resonate not only in living cells but also in human minds now striving to understand the meaning of it all.

Living beings are dramatic not because they function in obedience to physical and chemical laws but because they function in accordance with what Polanyi calls the "logic of achievement."[8] Living beings are capable of tragically failing because, first of all, they strive to be successful in achieving goals of various sorts. Sometimes they succeed, but at other times they fail. By contrast, astronomical, physical, and chemical processes, considered apart from living processes, neither strive, nor achieve, nor aim, nor fail. Because of life, however, the universe is dramatic. Let us now examine three ways of reading the drama of life.

Archaeonomy. Nowadays, the archaeonomic reading of life is standard. This academically approved way of looking at life ignores the drama altogether. Archaeonomic naturalism by definition blinds itself to the logic of achievement that distinguishes life from nonlife. Archaeonomy's disciples try to understand life by leaving out what is most distinctively "living" about organisms: their capacity for anticipation. Ironically, the human devotees of archaeonomy themselves anticipate specific goals in their cognitive lives. They *strive*—but so far have failed—to reduce organisms and cells completely to nonliving elements and routines. Clad in the armor of archaeonomy, these living (human) beings *aim* to achieve a comprehensively physicalist understanding of life. In doing so, however, they fail to notice that they, too, are living beings, striving in this case to reduce life to nonlife.

The archaeonomic assumption that lifelessness is the fundamental—and most intelligible—state of cosmic existence may seem understandable. Recall here the thirty volumes representing our 13.8 billion-year-old universe as I pictured it in Chapter 3. In that portrayal of deep time the first twenty-one volumes are completely devoid of living organisms. This immense temporal span of cosmic languishing may seem to support the view that dead matter is the mother of all things. Taking that picture literally, archaeonomists endeavor to account for the later chapters of the cosmic story in terms of the lifelessness featured in the first twenty-one volumes. Since lifelessness dominated the cosmic landscape for so long, deadness may seem to be the defining quality of the whole cosmic story.

Archaeonomists aim to reduce life to lifelessness, at least in theory, and a good percentage of scientists and philosophers still agree with Francis Crick (1916–2004), who claimed that "the ultimate aim of the modern movement in biology is to explain all of life in terms of physics and chemistry."[9] Notice, though, how the dramatic logic of achievement infiltrates

Crick's claim. He and many other materialist scientists aim to reduce life to nonlife. This aim, however, presupposes that each scientist is a living subject who is striving intentionally to bring about that result. Crick's associate James Watson (b. 1928), codiscoverer of the double helix formation of DNA, anticipates that "life will be completely understood in terms of the coordinated interactions of large and small molecules."[10] Biology departments, if this claim ever turns out to be literally true, should expect to be swallowed up by chemistry and physics departments. If Crick and Watson are correct, we cannot help asking whether biology departments in the future will justifiably claim any academic independence, since they will be housing specialists who, it seems, are striving mightily to do away with their own discipline.

The archaeonomic stance assumes that whatever happens in the cosmic story can be nothing more than a reshuffling of lifeless physical units. In carrying the search for intelligibility back in time to the original subatomic stage of cosmic history, a purely archaeonomic reading identifies the universe not with an ongoing drama of awakening, as I am doing in this book, but with the inanimate physical stuff out of which life accidentally arose. Archaeonomists almost deliberately close their eyes to the dramatic quality of life without which their own minds, now striving for right understanding, could never have achieved any success.[11]

Seduced by the lifeless look of the early universe, archaeonomic materialists equate true being with what has been. They imply, therefore, that what will be, as the cosmos continues its awakening, can be nothing more in the final analysis than the physical equivalent of what is chronologically original and seemingly fundamental—namely, a lifeless sea of physical elements guided by impersonal laws. Instead of wondering about the drama of a universe that has made its journey from lifelessness to life, and then from life to the existence of minds such as their own, archaeonomists are content to interpret preliving cosmic immensity (space, time, and vast numbers of lifeless atoms) as evidence of the essential lifelessness of being.[12] Archaeonomically speaking, the immensities of space, time, and numbers do not make up a generous matrix for life but are proof of the universe's deadly indifference to it. The long-drawn-out temporal pace of the physical processes leading to life, moreover, appears to provide further evidence for Monod's claim that life does not really belong to the cosmos.

Today, many scientific thinkers are not as merciless in their materialism as Monod was. Some allow that matter is inherently open to "emergence." That is, they acknowledge—in an almost analogical spirit—that new organizing principles occasionally drop into the material world, providing unprecedented formal arrangements of physical reality. Still, most scientific attempts to make sense of emergence are, in my opinion, far from abandoning archaeonomic habits of thought. There has yet to appear a widely acceptable worldview or metaphysics among scientists and philosophers that can render the phenomenon of emergence fully intelligible. To that end, I am aiming to present an anticipatory vision of the universe as a reasonable alternative. Readers can judge whether, in the final analysis, I succeed or fail.

A purely archaeonomic reading of nature fails, as I see it, because it is logically self-subverting. It reduces to mindlessness the very minds of those who strive to hold on to their materialist worldview. Yet, in spite of this fatal flaw, archaeonomic materialism even in the twenty-first century is generally considered the royal road to right understanding. I doubt, however, that even the most reductionist scientists really believe that their minds are nothing more than mindless stuff. If they did, there would be no justifiable reason for the rest of us to pay any attention to them. Consequently, to make good sense of their own minds, and to justify the trust they have in their own cognitive skills, they need to find a way of reading the universe other than archaeonomy. My proposal is that they try out the way of anticipation. First, though, let us consider the analogical reading of our life-bearing universe.

Analogy. Analogy has consoled religious people for centuries because it offers to connect their perishable existence to an eternal source of life outside of nature. The analogical stance assumes that life, like human souls, belongs to eternity more than to time. Life appears precariously on Earth, but according to the analogical stance, its instability proves that it does not belong fully to the physical universe. Life is tied only tenuously to the passage of time, which itself seems unreal. The final destiny of life is to be released eventually from its exile in matter so that it may return to its place of origin, eternity itself. Life belongs properly not to a cosmic story but to a supernatural holy land where time, perishing, and death do not exist.

In the philosophical parlance of analogy, life is irreducible to, and "ontologically distinct" from, pure matter. The intuition that life is discontinuous with mere matter is an essential part of what has come to be known as the Perennial Philosophy. This is a body of teachings about nature and God that prescientific philosophical and religious traditions have held in common for centuries.[13] Perennialism still has numerous adherents. Foremost among its teachings is that there exists a single ultimate reality, a timeless, eternally living, and transcendent source of the world's being—what Christians call God. This divine reality is the font of all finite being. It is more real and more alive than anything temporal. Matter is the lowest and least important level in the perennialist hierarchy of being. Above pure matter lie the higher levels of life, mind, and spirit. In the Nicene Creed, for example, the "Holy Spirit" who "proceeds from the Father and the Son" is identified as "the Lord, the giver of life." The more real and the more important levels of being are immaterial, since matter, because it is contaminated by time, is the lowest level of all.

Today, of course, science has demonstrated that life came into the universe long after inanimate stars, atoms, and molecules. Even the most entrenched perennialist must admit that lifeless matter, terrestrially understood, has at least chronological priority over life. Didn't life, historically, come from matter? So how can life be more important than its progenitor? On what basis can perennialists still affirm the ontological superiority of life over matter, especially since natural history tells us that matter gave rise to life gradually, over a long period of time? If we look at the whole cosmic story, isn't mindless matter the mother of all things? Aren't the atomists and archaeonomic materialists justified in reducing life to mere matter, and biology to chemistry and physics?

The very essence of archaeonomy is its assumption that chronological priority is equivalent to ontological priority. That is, physical stuff is more real than life or thought because it came earlier and lasts longer. By breaking living beings down into their irreducible lifeless physical components, the archaeonomic stance claims in effect that what is most real about our universe is its opening stage of particulate material dispersal. All the complexity that comes later—including living and thinking beings—is unreal in comparison with the lifeless physical units that came first.

By carrying our minds figuratively back to the beginning of time, however, pure archaeonomy leads us not to coherence but to physical de-coherence. Archaeonomy eventually leaves our thoughts stranded in a swarm of primitive subatomic units.[14] By contrast, the analogical approach teaches us to value life as something special. Life is more real and more important than mere matter. Chronological priority does not entail ontological superiority.

How, then, does life come about? Analogically speaking, life is the result of eternity intersecting with time, not the effect of some vague cosmic alchemy that mysteriously transforms deadness into aliveness. Living beings are to be valued because in all their fragility they participate in—and remind us of—what is timeless, otherworldly, and indestructible. Devotees of analogy cherish organic life because it puts us somehow in touch with the eternal life of God. As far as humans are concerned, each of us is thought to have an immortal soul. The soul is a special form of life that gives us a special connection to God even in the midst of time. The human soul allows us to share more fully than any other terrestrial beings in the ever-living goodness and beauty from which time is a departure.[15]

To analogy, therefore, life is special because it is ecstatic, not because it is dramatic. Life is breathed into a lifeless cosmic setting from up above, giving matter a touch of eternity. Eventually life returns to its timeless divine source on high. Analogy is not impressed by contemporary scientific claims that life, in its many manifestations, is merely the outcome of a long and convoluted story of physical transformations going on in deep time. Its devotees generally ignore as inconsequential the unfathomably long natural history that gave rise to life. Analogy values life, not because it is the outcome of a story going on in deep time, but because it partakes presently of an eternal fullness that lies altogether outside of time.

The analogical valuing of life arose in human consciousness long before contemporary scientific discoveries demonstrated that complex physical and chemical transformations, along with the immensity of space and time, are essential to the arrival and development of life. Analogically minded Christians today, as a result of their religious education and perhaps classical theological training, have difficulty experiencing a strong personal connection to the long cosmic story from which life has emerged historically. Scholars among them continue to look longingly toward theologies that never had to face fully the fact of deep time. A good example of this nostalgia is the nineteenth- and twentieth-century Roman Catholic return to medieval thought as the most appropriate intellectual framework for a sacramental appreciation of life. Defending itself against modern atomistic materialism, Catholicism officially sanctioned the thirteenth-century theological system of Thomas Aquinas as an alternative to, and defense against, the materialist reading of life.

Aquinas's affirmation of the classical hierarchy of being and the transcendence of God provides a peaceful spiritual and intellectual haven for reaffirming the value of life, and this is partly why Thomism is still considered the official philosophical setting for Catholic interpretations of the Nicene Creed. The Thomist medieval intellectual system allows for the incarnation of God and affirms the association of God with time. It also respects the temporal material world. Thomism, I should note in passing, was the main intellectual framework of my own early theological training, and I have a qualified fondness for it even today. But does it love time enough? And can its analogical vision be stretched far enough to capture the dramatic quality of life? I doubt that it can, for reasons that I am spelling out in this book.

After Darwin, during the late nineteenth and early twentieth centuries, the ghost of analogy—still floating around in the minds of a few evolutionary thinkers—sometimes morphed into a philosophy of life known as vitalism (from the Latin word for "life," *vita*). Vitalism hypothesizes that life is the result of an immaterial, presumably divine, force entering into nature from beyond. Life exists because an infusion of special energy into matter has caused cells and living organisms to spring forth in some mysterious way out of lifeless matter. The French philosopher Henri Bergson (1859–1941), the most famous of modern vitalists, speculated that a nonphysical impetus—an *élan vital*—is required to give rise to life. This vital energy at work in nature, he claimed, is accessible only to intuition, not to scientific observation or analysis. It is completely out of the range of the archaeonomic reading of nature.[16]

During the first half of the twentieth century vitalism was attractive to philosophers and scientists who were disenchanted with the prevalent materialist interpretations of nature, life, and evolution. Bergson's vitalism appealed to analogical theologians because it considered life more real than matter. Moreover, Bergson took the passage of time more seriously than did Einstein and other scientists who considered our human sense of duration to be a subjective fiction. Up until the more recent molecular revolution in biology, in which the archaeonomic approach began to dominate the field, Bergson's philosophy seemed quite tenable intellectually.

No longer, however, is that the case. Vitalism is an understandable reaction to archaeonomic materialism, but it now seems insufficient both scientifically and theologically. Scientifically, the problem is that, like other versions of the analogical stance, vitalism fails to notice how the three cosmic immensities—time, space, and complexity—are woven narratively into the origin, development, and survival of life. And, theologically, vitalism compromises too much with archaeonomic materialism by divesting the earlier ages of cosmic history of their dramatic complicity in the emergence of life.[17]

Vitalists, though sometimes notionally aware of deep time, remain stuck in analogy to the extent that they fail to acknowledge the seamless dramatic connection of life to the physical universe. The vitalist instinct is to set life in sharp contrast to lifeless matter. Vitalism turns life into an interruption of nature rather than a dramatic outcome of a cosmic transformation. Instead of reading the whole history of matter as internal to the drama of life, as I am doing, vitalism no less than materialism fails to appreciate the narrative connection of life to the whole cosmic story.

Even though Bergson rightly affirmed the reality of irreversible time, in his disputes with Einstein he unfortunately continued subtly to separate life and time from matter, as in earlier versions of analogy. It might seem at first—as it did to some of Bergson's early supporters—that a vitalist philosophy of nature is perfectly compatible with Catholic religious beliefs. Yet by placing tension and perhaps opposition between life and matter, vitalism seems insufficiently incarnational, theologically speaking. In my opinion, it fails to connect God as intimately to time and the history of matter as the Nicene Creed in principle allows.

Anticipation. A vitalist version of analogical reasoning, or something vaguely akin to vitalism, is still favored by many scientifically educated religious believers. Vitalism is spiritually and morally attractive because it upholds the traditional theological emphasis on the special value of life. And it is intellectually attractive because it does not seem to conflict with science so much as with materialist interpretations of science.

After Einstein, a vitalist reading of life is not required for the reconciliation of science with theology. An appropriate place to locate the divine source of life is not outside of time but in the region of the not-yet. If, as I have been arguing, the sense of God requires our taking an anticipatory stance, so also does our awareness of life. Vitalism, however, adheres too closely to the analogical reading of nature, and by doing so it fails to appreciate fully the dramatic significance of the preliving stages of natural history. Vitalists do not seem to notice how tightly the immensities of space, time, and complexity are linked to the drama of life's emergence. By explicitly or tacitly assuming a division of life from matter, vitalism is forced to account for the exceptional properties of life by postulating the existence of a nonphysical kind of causation whose supernatural origin places life beyond scientific scrutiny.[18]

In both the archaeonomic and the analogical readings of the cosmic story life arises in spite of, rather than because of, the fundamental features of the physical universe. In neither of these two readings does life fully belong to the physical universe. By contrast, an anticipatory reading of nature thinks of life as completely continuous—dramatically speaking—with the nonliving epochs of the cosmic story. To anticipation, the universe's ultimate principle of coherence lies up ahead, in the future, whereas to archaeonomy it lies in the remotest cosmic past. To archaeonomy the universe's spatial, temporal, and numerical immensities are indicative of the universe's indifference to life. To anticipation, however, the late and local eruption of life is an occurrence that retrospectively gives dramatic meaning to the antecedent cosmic immensities.

To anticipation, the issue is not whether or how the three cosmic immensities make life intelligible. The point is that the arrival of life—for which the universe had to wait—now gives a surprising dramatic intelligibility to the cosmic immensities. To archaeonomists, the physical universe overshadows our brief lives and crushes us without leaving a trace. The anticipatory stance, on the other hand, is impressed by a universe wherein the origin and development of life (and mind) finally render the cosmic immensities dramatically intelligible to those who are willing to wait. Analogy also is too impatient to grasp or appreciate the narrative meaning of our three cosmic immensities. It turns abruptly toward eternity to avoid inquiring about the meaning of deep time. Vitalism, in its compromise with materialism, views life as something that happens in spite of, rather than because of, the cosmic immensities.

The cosmic immensities laid bare by relativity and Big Bang cosmology may seem at first to make the cosmic journey toward life more circuitous than our magical mindsets prefer. Anticipation, however, is not surprised to discover in the universe a wildly adventurous itinerary involving innumerable experiments, near misses, periods of suspense, and even dead ends prior to life's arrival. The three immensities give a narrative depth to organic life that neither materialism nor vitalism can savor. By failing to acknowledge the dramatic character of time's passage, archaeonomy cannot even see life, let alone value it.

Analogical piety typically ignores and belittles the creative role of deep time. It takes the shortcut of connecting each living organism vertically to a timeless sphere of being that is supposed to rescue lives and souls from irreversible time and the dramatic current of cosmic duration. An anticipatory stance, in reaction, has no room for theologies or philosophies of nature that try to calm our spiritual restlessness by taking analogical flights into a timeless beyond. Anticipation fosters a stance of expectancy that links life here and now to a future blossoming in time. It seeks the ground of life's being and value by looking toward the future, not the eternal present or the fixed past. To anticipation, the existence of life is already an exceptionally wondrous arrival of the future. Looking to the future—to the not-yet—to discover the ground and essence of life is contrary to the assumptions of both archaeonomy and analogy, but it is not contrary to the demands of either science or faith.

The archaeonomic approach sees only absurdity in the fact that the cosmos "wasted" billions of years, squandered wide tracts of empty space, and engaged in a profligate manufacturing of hydrocarbon molecules before letting life come into the cosmic story. Archaeonomists such as Monod, as I have been pointing out, are not accustomed to waiting, so they see little, if any, meaning in the drama of cosmic duration. Archaeonomists assume that the universe would make more sense if it had been better engineered initially. Anticipation, on the other hand, with its sense of a universe still being born, is prepared to wait for the meaning of matter and time to emerge in the future.

To anticipation, finally, the universe is convergent rather than divergent. The general movement of the cosmic story is from dispersal toward unity. According to Teilhard, Bergson misinterprets evolution as "a radiation that spreads out from a central source." By contrast, Teilhard's anticipatory approach interprets the evolution of life, at least overall, as a movement from multiplicity toward unity. It does not always seem that way, as Bergson's impressions of life's divergence indicate. But Teilhard replies:

> It is not to be wondered at that all the roads that life tries in order to effect the synthesis of the multiple are not equally profitable. Drawn by the same unifying force, beings choose different roads, some more direct, some less. Some of them come to grief in their attempts, or take a retrograde path. As the elements . . . make their way together towards their common goal, they suffer a disintegration—what one might call an unraveling—of their fibres. That is why so many phyla break away and

> give the illusion of a divergent evolution. This fragmentation is superficial and secondary. Basically the whole of the world's psychism gravitates toward a single centre.[19]

Life in a Multiverse

How would archaeonomically inclined scientists and philosophers respond to what I have proposed above? Astrophysics, they would now have to agree, has demonstrated that the Big Bang universe from its very beginning has been endowed with physical characteristics and mathematical values precisely appropriate to the eventual coming of life into the cosmos. Given the physical properties of the early Big Bang universe, they would agree, life on Earth seems not so improbable as it did to Jacques Monod. However, to secure their belief that the existence of life does not threaten a materialist picture of the universe, cosmologists may try to reestablish the reign of archaeonomy by hypothesizing a multiplicity of lifeless universes. Then, if there are enough of these—that is, if there is a multiverse—lurking in the background, and if most of the separate worlds do not have the right physical properties to support the existence of life, nonetheless—in an unimaginable plurality of worlds—the immensity of numbers, time, and space increases the probability that life might accidentally pop up somewhere.[20] In such a broad statistical setting the existence of life as we know it would still be a pointless blip that does not compel us to abandon the archaeonomic intuition of nature's overall meaninglessness.

The anticipatory stance, in response, would have no difficulty in principle with the idea of a multiverse. Indeed, the existence of many worlds may even be considered likely if creation is grounded in a font of divine superabundance. If we take an anticipatory rather than an archaeonomic stance, the eventual emergence of a single life-bearing universe may even bring dramatic intelligibility to a supposed multiverse. Just as the emergence of the first living cell on Earth brings dramatic intelligibility to our three antecedent cosmic immensities, so also, in principle, may the arrival of life on only one planet bring dramatic intelligibility to the hypothetical background of multiple universes. The immensity of time, space, and large numbers of things increases the probability that somewhere, at some time, the conditions for life will arrive. But we have to wait for that to happen. In an anticipatory universe it is only after we have figuratively waited for and witnessed the arrival of life that we can look back and make dramatic sense of the multiverse.

Trying to explain the existence of life on Earth solely by multiplying universes leads our minds not forward toward coherence but backward in the direction of de-coherence. From an anticipatory perspective, however, the relatively recent arrival of life—even if only at a single location in a vast multiverse—can in principle bring narrative coherence to the whole plurality of antecedent universes. But we have to wait. Archaeonomic impatience leads cosmologists to multiply lifeless worlds imaginatively to explain the existence of life on one lonely planet. Anticipation finds that the existence of life—even if only on Earth—can give dramatic unity and intelligibility to a whole array of worlds in a multiverse.

Summary

An anticipatory understanding of life does not have to compete in any way with the scientific research going on with respect to the physical and chemical origin and makeup of living beings on Earth. When the Nicene Creed refers to the Holy Spirit as "the Lord, the giver of life," it is speaking dramatically rather than scientifically. When scientifically educated Christians recite the Creed, they do not have to worry about any conflict with what they have learned from the classroom or laboratory about the spontaneous beginning of life, about its molecular makeup, or about the long and gradual evolution of species. Nothing in science contradicts the anticipatory sense that the emergence of life brings dramatic new meaning to the whole story of nature.

Life is especially dramatic because organisms have the capacity to strive, succeed, and fail, thus introducing the notes of struggle and tragedy into the universe. But since the existence of life is tied narratively into the large-scale features of nature, the epoch of life on Earth gives a dramatic coherence to the entire cosmic story. And if it takes many worlds to make a life-bearing universe, then the existence of life anywhere gives dramatic intelligibility to the multiverse as well.

Vitalism's analogical indifference to the nuanced history of matter leads it to interpret life, not as an extension of the story of matter, but as a kind of disturbance of it. Vitalists, even when they are aware of contemporary cosmology, are likely to consider the cosmic immensities and the universe's tumultuous transformations extrinsic to life. An anticipatory reading, on the other hand, looks at life for its dramatic meaning. Anticipation fixes its attention, patiently and hopefully, on what is coming into the present from the arena of the not-yet. Life itself is a major installment in the dramatic arrival of the future.

EIGHT

Thought

> This universe of ours is something perfect and is responsive to the rational striving for knowledge.
> —Albert Einstein

> I have found no better expression than "religious" for confidence in the rational nature of reality insofar as it is accessible to human reason. Whenever this feeling is absent, science degenerates into uninspired empiricism.
> —Albert Einstein

> The effort to strive for truth has to precede all other efforts.
> —Albert Einstein

BY 1931, EINSTEIN HAD reluctantly come to agree that the universe depicted in his field equations was not standing still. Mathematical interpretations of relativity by Alexander Friedman (1888–1925), Willem de Sitter (1872–1944), and the Belgian physicist and Catholic priest Georges Lemaître depicted a changing universe. New observations of galactic movements by the American astronomer Edwin Hubble demonstrated, by measuring the varying lengths of light waves arriving from far-off galaxies, that the universe is still undergoing an overall expansion. Lemaître had been right in surmising that the universe started off as an immensely dense "primeval atom" long ago, that it moves irreversibly from past to future, and that it is a process still under way.[1] Even after giving consent to this new cosmic picture, however, Einstein did not think of the universe as dramatic, and he never seems to have asked what it means for our understanding of the cosmos that it has given birth to minds. Nor did he wonder why these minds—not least his own—are now insatiably searching for more meaning and deeper truth.

The universe has produced a lot of impressive outcomes over the past 13.8 billion years, but surely none more significant than the human mind. With the emergence of mind, the universe has now become capable of not only experiencing, but also understanding, reflecting, knowing, and deciding. We may refer to this whole suite of mental acts simply as thought. It is especially in the phenomenon of thought that the universe is now awakening to more meaning, truth, goodness, and beauty. Why, then, do so many thoughtful people today confidently claim that the universe is pointless? If purpose means "bringing about something of self-evident importance," then the universe can be considered pointless only if we forget that it has been manufacturing minds that are responsive to the indestructible values of meaning and truth: minds like yours, mine, and Einstein's.

If modern science has taught us anything, it is that thought is inseparable from the cosmos. The physical universe, having slumbered for billions of years, has gradually become alive, sentient, and conscious. In human beings it has become not only capable of experiencing, understanding, and knowing but also conscious that it is conscious. What makes us humans distinctive in nature is not just our upright posture, our talent for toolmaking, or our patterns of moral behavior. It is not even our capacity for experiencing and understanding. Mostly, it is our remarkable ability to know that we know.[2]

After the origin of life on Earth 3.7 billion years ago, nature gave birth to beings that became increasingly capable of inner experience. In the animal kingdom nature has long included many kinds and degrees of subjectivity—that is, the capacity to register and remember experiences. Recently, in centers of experience known as human beings, the natural world has become reflectively conscious of itself. What does it mean for our understanding of the universe that it has now given birth to reflective thought?

Archaeonomy. The first of our three readings cannot accept the idea that the birth of thought makes any significant difference in how to understand the universe. Archaeonomy's proponents claim that mind, along with the rest of nature, is ultimately reducible to mindless material stuff.

Mind comes into the universe by way of a series of blind elemental movements and, later on, evolutionary accidents and adaptations, starting in the original state of senseless subatomic plasma billions of years ago. Since archaeonomists reduce everything to original mindlessness, they need to tell us how the complex phenomenon of thought could emerge from such a simple and lowly ancestry. Only if there is plenty of time, they answer, can an initially mindless universe gradually grind out brains and minds capable of thought. In making this claim, however, they cannot help trusting in the capacity of their own minds to find intelligible explanations and discover truth. A good question to ask them, then, is whether this trust they have in their minds is justifiable. Why should they, or we, trust minds that are reducible to mindless matter?

On the one hand, archaeonomic materialists tell us that mindless matter is all there really is and that all minds are reducible to insensate physical stuff and the impersonal laws of nature. On the other hand, in the very act of making this claim they have to trust in the integrity of their own mental functioning. Even though their method of inquiry reduces life and mind to lifeless and mindless elements, the minds that are doing this reducing seem to exempt themselves from being part of the atomized unconsciousness into which their worldview has decomposed the universe.

It is fascinating that materialist scientists place such extraordinary trust in their own minds, and that they do so even more confidently than almost any other body of contemporary thinkers. Even the most extreme physicalists among contemporary cognitive scientists secretly doubt that their own minds are reducible to the mindlessness that prevailed in the early universe. The irony is that their archaeonomic reading of nature provides every reason to doubt that their minds can be trusted. Yet the more they insist on the ultimate mindlessness of nature, the more they function with unquestioning faith in their own exceptional cognitive skills.

The arch-archaeonomic philosopher Daniel Dennett, for example, revels in his own exceptional mental agility. He is highly confident of his own cognitive skills. We can witness this confidence from book titles such as *Consciousness Explained,* a work in which the author exclaims victoriously that "there is only one sort of stuff, namely matter—the physical stuff of physics, chemistry, and physiology—and the mind is somehow nothing but a physical phenomenon."[3]

That claim, however, would have to apply to Dennett's own mind, too. His own thoughts must emanate from a realm of being that he considers to be just "physical stuff." But if Dennett's mind is "nothing but" a physical phenomenon, how can he logically justify the monumental trust he has in that singular mind's capacity for right understanding? And why should his readers respect his thoughts, as he expects us to do, if they spring from an abyss of mindlessness?

I am not at all implying here that Dennett should not have confidence in his cognitive ability. He should. He simply needs to find a way to justify that confidence, since a materialist worldview will not do the job. Logically speaking, Dennett's cognitive confidence is undermined by his materialist claim that all minds are nothing but physical stuff. There is no getting around the logical incoherence of this archaeonomic stance. Yet in philosophical and scientific circles today archaeonomy continues to flourish, and Dennett is by no means alone in believing that stance to be beyond criticism.

Owen Flanagan is another respected archaeonomic thinker. This Duke University professor declares without the slightest hesitation that the objective of academic philosophy today is to make the world safe for (archaeonomic) naturalism. He fails to notice, however, that a purely materialist worldview logically subverts his own cognitive confidence. In the very act of claiming ontologically that human intelligence has its ultimate explanation in the combination of mindless matter, endless spans of time, and impersonal physical laws, Flanagan fails to give us any good reasons why we should pay attention to the thoughts that spring from his mind.[4]

It is no less illogical, moreover, that so many archaeonomic materialists like Dennett and Flanagan today explicitly embrace Darwinian evolution as the "ultimate" biological explanation of human intellectual functioning. When they "explain" mind as an evolutionary adaptation, they are forced to agree that human minds—and this would have to include their own—are the outcome of an aimless series of random events sculpted by a mindless mechanism known as natural selection. But, again, they fail to tell us why any mind that is reducible to mindlessness can be trusted. We know from biology that minds are adaptive, but how do we know that they are able to gift us with right understanding.

"Rationality is not the gold standard against which all other forms of thought are to be judged," David Sloan Wilson boldly asserts. Rather, he continues, "[evolutionary] adaptation is the gold standard against which rationality must be judged."[5] What does he mean by this? On the one hand, Wilson, a justly respected evolutionary scientist, is telling us that human mental functioning is the outcome of a long terrestrial process of mindless Darwinian adaptation. Yet in writing and publishing his books, he is asking his readers to trust that his own mental functioning is leading us to truth and not merely adapting to environmental constraints. Wilson may attribute his cognitive confidence to societal, familial, and educational formation, but the gold standard of rationality, he claims, has its ultimate explanation in mindless evolutionary adaptation.

Even though Wilson's evolutionary archaeonomy allegedly accounts for the attributes of minds in terms of the blind mechanism of mindless adaptation and the aimless passage of time, he considers his own mind—the outcome of an ultimately mindless process—to be of sufficient integrity to put us in touch with the real world. Why should we consent to any of Wilson's claims if all mental functioning, including his own, is ultimately a matter of adapting and surviving? After all, one of life's most adaptive and survivable traits is the capacity to deceive, as evolutionary naturalists have often observed. How do I know that Wilson is not deceiving me in the process of his own mind's adapting to the universe?

What I am questioning here is not evolutionary biology—which I enthusiastically accept—but the reasonableness of archaeonomic readings of nature and thought. How do human minds, including those of reductionist scientists, acquire the power to arrive at truth? Given a materialist worldview, how can archaeonomic thinkers such as Dennett, Flanagan, and Wilson justify the trust they have in their own mental capacity, especially since they claim formally that the roots of rationality are planted in the deep soil of pure mindlessness? Shouldn't such internal self-contradictions be enough to make reasonable people skeptical of the archaeonomic stance taken by so many contemporary scientific thinkers? Furthermore, to claim, as some evolutionists do, that our capacity for thought is a by-product rather than a direct outcome of mindless evolutionary adaptation, does nothing to establish its cognitive trustworthiness either.

The philosopher Thomas Nagel has recently exposed the intellectual incoherence of all purely materialist theories of mind. His well-crafted critique has been met with impassioned scorn by materialist philosophers of mind, who remain supremely confident of their own capacity to reach right understanding. Nagel, unfortunately, lacks an alternative, nonmaterialist worldview, one capable of refuting archaeonomic materialism without giving the appearance of rejecting the solid scientific foundations of evolutionary biology and contemporary cosmology.[6] As a reasonable alternative we may consider an anticipatory understanding of thought as outlined below.

Even the most materialist of contemporary thinkers, it appears, do not really believe that their own minds are nothing but physical phenomena. This, I believe, is an indication that they are still analogical Platonists at heart. To guarantee the integrity of their own cognitive operations archaeonomists tacitly uproot their minds temporarily from their allegedly mindless physical matrix and move them stealthily to an immaterial haven of timelessness. Even while their archaeonomic method of inquiry is in the act of reducing life and mind to lifeless and mindless atoms, the minds doing this reducing are hovering supernaturally above the mindlessness into which they profess to have resolved the whole of nature.

Analogy. Ancient myths and philosophies often attempted to justify human cognition by separating minds from the mindless material universe that was temporarily imprisoning them. Minds, according to Orphic mythology and later Platonic metaphysics, are unworldly intruders into a foreign, inhospitable material domain. Consciousness, according to the ancestral myths that gave rise to the analogical stance, is a precious but homeless spark of light that has fallen into the world of matter like a meteor from the sky. Only because minds were thought to have been born in heaven above and not in matter below could they be trusted to lead our thoughts back to the timeless realm of truth and goodness from which they were temporarily estranged.

The mythic teaching that mind is not part of nature has persisted not only in medieval theology but also—unconsciously—in modern materialist thought. Though formally denying it, archaeonomic materialists are still captive to ancient myths that separated spirit from matter, soul from body, and mind from nature. Dividing reality into two separate spheres is known as dualism, an ancient habit of thought that came back to roost in a new way with the philosopher René Descartes (1596–1650) at the beginning of the modern age. Dualism still lives on, hiddenly, even in the most materialist instances of contemporary thought.

It is not surprising that archaeonomists risk midnight excursions into the forbidden world of analogy. The promise of analogy, after all, is that by connecting our minds to eternity they are protected from being dragged down into the sludge of mere matter and the endless perishing that goes on in time. For centuries, the analogical vision gained philosophical traction because it provided a plausible answer to the question of why we may trust our minds. It is because our homeless minds participate—though imperfectly—in the timeless sphere of Truth-Itself that we may trust them to lead us to right understanding, though not without experiencing constant temptations to stray. Even though they formally deny it, archaeonomic materialists are still secretly tied to analogical mythology.

Einstein, too, fell back into analogical thinking in his patently religious cultivation of the timeless mystery of cosmic comprehensibility. His attraction to Spinoza reflected an ancient religious intuition that nature and mind can be simultaneously ennobled only by their association with eternity. Unlike Dennett, Flanagan, and Wilson, Einstein could at least give a reason for the trust he placed in his own thought processes. The awareness he had of his mind's native contact with the transcendent mystery of cosmic harmony and comprehensibility justified the confidence he had in his cognitive functioning. He was not so slavishly bound to materialism that he let his own mind blend into the incoherence to which the archaeonomic stance reduces both thought and the universe.

To the pure archaeonomic materialist, as we have seen, every present and future state of the universe, including the recent arrival of thought, is resolvable into a past series of mindless physical movements and elements going all the way back to the beginning of time. Archaeonomy, because of its morbid attraction to an original cosmic deadness, sees no story, no awakening, and hence no narrative coherence in nature. In the archaeonomic reading, formally speaking, everything in the universe, including the brilliant minds that have gifted us with science, must be nothing more in the final analysis than the outcome of a mindless sequence of physical states and blind evolutionary adaptations. Analogy is right to insist, then, that the archaeonomic reading of nature, logically speaking, cannot be the road to right understanding of either the universe or thought.

A serious quest for intelligibility, as I have noted, is equivalent to a search for coherence, but the archaeonomic reading leads our minds back in time to a state of cosmic de-coherence and then abandons them there. Traveling all the way back to the primordial cosmic state of subatomic dispersal, where nothing other than mindless elements reside, archaeonomists find themselves again and again at an explanatory dead end. No genuine coherence or intelligibility is to be found by digging up the mindless units and blind physical laws to which archaeonomic materialists formally attribute the status of true being.

Devotees of analogy admit that we humans have no good reason to trust our minds unless they have somehow made contact with an indestructible rightness. The problem is how to understand this connection in terms of recent scientific cosmology. Analogy professes to link our minds tightly to eternal meaning and truth, but in doing so, it fails to pay close attention to the irreversible temporal drama of nature that gave rise to those minds step by step. Analogy, including its vitalist versions, interprets mind as an interruption of matter and time, rather than as the outcome of a long, continuous natural history. Analogy, no less than archaeonomy, fails to acknowledge the irreversible, dramatic quality of time, and thus it fails to connect our minds satisfactorily to the whole long story of cosmic awakening.

Anticipation. Because Einstein's own thoughts flitted back and forth between archaeonomy and analogy, he failed to appreciate the long cosmic story within which we must now locate the birth and development of thought. As long as the cosmos seemed to be essentially unchanging, and as long as nature, in modern intellectual culture, appeared to be essentially mindless, thinking beings could not help feeling somewhat disconnected from nature. The mental functioning of human beings, in both the materialist and the Platonic interpretations, seems only loosely tied to the universe.

After Einstein, I submit, an anticipatory reading allows us to interpret our remarkable powers of thought as part of a longer cosmic awakening, one that has been going on continuously in time. Anticipation notices that the universe was already awakening in the earliest microseconds of its temporal existence. In the dawning of reflective human thought, then, we witness a dramatic eruption, rather than an interruption, of nature.

Our minds are not the product of an exceptional injection of some sort of thinking substance stored timelessly above the physical plane in an immaterial reservoir of perfect being. Nor are our admirable powers of thought coherently accounted for by the arbitrary claim that they are nothing but the result of a reshuffling of mindless atoms over billions of years. Rather, we are given sufficient reason to trust our minds if we interpret their existence as part of a continuous cosmic drama of awakening to indestructible rightness. Mind (or thought) *is* the universe in a relatively new and fragile stage of awakening to the horizon of infinite comprehensibility, truth, goodness, and beauty.

We do not, therefore, have to detach our minds analogically from nature to find a reason to trust them. Instead, we note that the cosmos itself is by definition, and not by accident, an awakening. So, we do not have to tear the human mind out of its physical and evolutionary matrix and place it outside of nature to justify our trust in it. Our minds are deeply embedded in the narrative of natural history. But the whole story of nature, as I have been arguing, is already a suspenseful anticipation of rightness. It is because the universe itself is an awakening, a response to the unquenchable light of infinite rightness, that our recently emergent minds can claim to be trustworthy. We may trust our thought processes, at least conditionally, to the extent that they are always being gently embraced and at times enraptured by the very rightness they anticipate. If we allow ourselves to be carried away by the intelligibility flickering up ahead, our minds are already exalted in a way that renders it reasonable for us to trust them.

Anticipation was already part of our universe long before reflective thought actually arrived in cosmic history. The origin of thought was an intensification of nature's pervasive anticipatory leaning, not an alien intruder. The universe, even from the beginning, could not have been accurately understood by a physicalist or purely geometrical accounting. So now, after making an analytical scientific journey back to the remotest subatomic cosmic past, archaeonomists should immediately turn around 180 degrees and cast their gaze in the direction of what is not-yet. Only by looking forward and waiting patiently for what is coming would they be able to justify the trust they have in their powers of thought.

Anticipation is not at all incompatible, I insist, with evolutionary biology and the new scientific story of mind's gradual physical emergence. While emphatically opposed to archaeonomic materialism, anticipation fully endorses scientific research, including Darwinian biology, neuroscience, and Big Bang cosmology. But by adopting an anticipatory rather than an archaeonomic stance, and by simultaneously embracing the fact that our minds are part of an awakening universe, we may justify our cognitive confidence without having to reject science in the slightest way. In the wake of Einsteinian relativity and contemporary cosmology the anticipatory perspective prepares us to be fully scientific, on the one hand, and gives us a worldview that can justify our cognitive self-confidence, on the other. All we need to do is learn to read the universe as an anticipatory story, as a drama that is always in the grasp of an elusive but indestructible rightness.

Admittedly, this anticipatory stance may be hard to adopt, since the archaeonomic and analogical ways of reading the natural world are still lodged so firmly in contemporary intellectual and theological habits of thought. Neither materialist monists nor analogical dualists have considered seriously and consistently that mind is part of an awakening universe. Learned thought in general has for centuries moved mostly back and forth between Democritus and Plato. Einstein's own opinions about science and religion still reflect this ambivalence. He never realized that relativity allows for an anticipatory reading of the universe, a stance that avoids the extremes of atomism and escapism.

Throughout his lifetime Einstein failed to appreciate the universe as something truly dramatic. His focus was not on matter's inherent narrativity but on nature's impersonal lawfulness, geometric structure, and underlying physical determinism. Like many other modern thinkers, he was more impressed by the universe's lawful regularity than by its dramatic adventurousness. His was a universe not of bounteousness but of balance, not of extravagance but of equivalence. He projected onto nature a juridical logic that automatically obscured the drama of cosmic awakening going on beneath the surface. And yet it was his own science that, after being mathematically tweaked, released the good news of a cosmos spilling forth narratively into an impressive, expansive drama of awakening.

In spite of that good news, archaeonomic materialists after Einstein still take the universe to be not only mindless but also meaningless. The Nobel Prize-winning physicist Steven Weinberg, for example, has declared that the more intelligible the universe has become scientifically, the more pointless it also seems dramatically.[7] The automatic adherence to cosmic pessimism on the part of most contemporary intellectuals stems, I believe, from the assumption that their own inquiring minds are not really part of the universe.

Weinberg's book *Dreams of a Final Theory,* for example, records his own poignant struggle to find dramatic meaning in the universe, as well as his sadness in failing to do so.[8] But his otherwise thoughtful reflections fail to acknowledge that his own quest for meaning is part of the cosmic story. Weinberg does not consider the possibility that the deepest roots of his own search for meaning lie in the anticipatory leaning of the universe that gave birth to his mind. Like most other scientific thinkers today, he quietly assumes that the operations of his own mind have nothing significant to tell us about the nature of the universe. To all of us, mind seems so different from impersonal nature that we are tempted to view it as something unnatural or supernatural. It strikes even most contemporary cognitive scientists that the mind's private, subjective experience is an unintelligible exception to the otherwise mindless state of the cosmos.

Only an anticipatory reading of nature, I am arguing, can keep us from going down the blind alleys of archaeonomy and analogy in our attempts to understand mind and thought. We may justifiably trust our minds only if we firmly reject archaeonomic materialism, but we can situate our minds realistically within the history of nature—as science requires—only if we also resist the analogical/Platonic temptation to ensconce them in a timeless, immaterial sphere apart from nature. Thought, as we look at it in an anticipatory way, is neither a mere reshuffling of atoms nor immaterial sparks of light floating into the cosmos from an immaterial world. Thought is fully part of nature, but nature itself is an awakening to the not-yet of indestructible rightness. The arrival of human thought in the cosmos is a significant episode in that larger drama.

Mind and Creed

After Einstein, interpreting the universe as a possible carrier of meaning requires that we distinguish between nature's physical, chemical, and mathematical comprehensibility, on the one hand, and its dramatic meaning, on the other. To grasp any dramatic meaning fermenting in the universe it is not necessary to look for physical interruptions or suspensions of nature's invariant rules. It is not inconceivable that something profoundly meaningful, invisible to a purely geometric calculus, is still going on beneath the surface.

The archaeonomic stance is right to highlight the continuity that all present states of nature have with the physical past. Classical analogical thought, for its part, is right to emphasize that something about human minds makes them stand out from their physical environment. The anticipatory stance claims that our minds' exceptional capacity for thought is not due either to a blind reshuffling of primeval atoms nor to their belonging to a timeless nonmaterial world but to the arrival of meaning, goodness, and truth from up ahead. The unbroken physical and chemical continuum of causes in nature does not rule out discontinuity at the level of dramatic transitions.

The entries of living organisms and conscious minds into cosmic history were momentous narrative leaps. But narrative departures and transitions in natural history do not require the breaking of any scientific laws. The analogical stance, whether that of Plato, medieval theology, vitalism, or modern mathematical physics, fails to register the fact that the universe has always had, in addition to its (timeless) geometric structure, an essentially temporal, dramatic way of existing and that each human mind is fully part of the larger unfinished cosmic drama of awakening. The anticipatory stance locates the phenomenon of thought at the forward edge of a temporal universe that over the long haul has been awakening—not without the drama caused by chance, setbacks, and long periods of silence—to fuller being and deeper meaning up ahead in the sphere of the not-yet.

In the layered reading of nature that I am presenting in this book, what seem to be breakthroughs at the level of the universe's dramatic development need not show up in any visible bending of the rules of nature at the level of physical and chemical routines. What is remarkable dramatically may seem quite unremarkable physically and mathematically. Biologically and neurologically your own mind, for example, has emerged in the history of nature without breaking any physical, chemical, or evolutionary rules. Yet the arrival of your mind, and that of others, gives the cosmic story a whole new dramatic twist invisible to the physical sciences. The content of your thought processes in the present moment could not have been specified beforehand by an eternity of synaptic analysis. Dramatically speaking, however, the existence and activity of your mind are evidence that the universe is undergoing a transformation whose depth and meaning can be captured only by reading the course of events in some other way than that of the natural sciences.[9]

For now, you may still suspect, along with other thoughtful people, that your cognitive life is so different from what goes on in rocks and rivers that it somehow lifts you out of the natural world altogether. Your mental activity may seem to be such a sharply delineated exception to natural processes that you cannot help assuming that it belongs to some other metaphysical sphere than that of the natural world. You are not alone. Dualism, whether explicit or implicit, is a perpetual temptation. Philosophers, theologians, and even scientists have often thought of mind as somehow separate from nature. But I believe that our attraction to the idea of a crisp division between mind and nature is the result of a failure to grasp that the universe is a single story that can be read at different levels of depth. Thought, from one point of view, is the outcome of a causally continuous physical process, but from another—equally legitimate—point of view it is part of the universe's dramatic awakening to incorruptible meaning, truth, goodness, and beauty.

How to translate the subjective experience of feeling and thinking into the objectifying categories of cognitive science is known as the "hard problem." Expressed in terms of the present book, the hard problem is this: How can the archaeonomic understanding of human brains—a stance taken by most scientists and philosophers of mind—illuminate or connect with the inward experience of being an actual feeling and thinking subject? The problem disappears, however, once we take the anticipatory stance. As I see it, the universe itself is the primary locus of awakening, and our human mental experience is an intensification of a long and yet unfinished cosmic awakening. There are no pure subjects separate from the cosmos. Subjectivity *is* the cosmos in the process of newly awakening.

In contemporary thought our three main ways of reading the universe sometimes intermingle, often in contradictory ways, but the figures of Democritus, Plato, and Abraham loiter in the background. Einstein followed Democritus by reducing nature to a multiplicity of elemental units controlled by impersonal physical laws. At the same time, he read the universe analogically by reverencing the timeless mathematical forms lurking beneath all appearances. But neither his archaeonomic nor his analogical angle of vision could let him feel the pulse of the dramatic awakening going on in every stage of cosmic history.

It is from inside an anticipatory vision of the universe, I am suggesting, that Christians may now recite the Creed that has shaped the content of their faith across the centuries. A truly anticipatory reading of the Creed, however, has yet to settle influentially into our theological and religious sensibilities. Analogy continues to inform most Christian spirituality, while archaeonomy remains the dominant intellectual force in secular culture and academia. Since Einstein's own reading of nature bounced back and forth between the archaeonomic and the analogical, he was unprepared to look at nature and mind in an anticipatory way. Nor, I believe, are most other scientists, philosophers, and classical theologians.

Influenced by the opening chapter of the Gospel of John, the Nicene Creed professes that Jesus, the incarnate Word, victim of human malice and subjected to death by execution, is also "Light from Light, true God from true God." In the person of Jesus, Christians believe, life has achieved a decisive victory over death. And because the whole universe is linked to an incarnate divine rightness, Christians trust that matter can never be dissolved into nothingness, nor time into timelessness. Even if the cosmos eventually collapses energetically, as seems likely, the moments of time that make up the cosmic story will have kept adding up narratively, accumulating ceaselessly in the compassionate and restorative memory of God.[10] Christians may thus hope that every episode of the cosmic story is saved from absolute death and destined for transformation in the everlasting source of light to which the universe is now awakening dramatically.

Summary

The phenomenon of thought, from the perspective of an anticipatory cosmology, is a relatively recent intensification of what has always been a drama of awakening. For all we know, there may be many kinds and degrees of awakening in the universe — or multiverse. In any case, the cosmic awakening is not a set of events confined to the brains of an exceptional terrestrially bound species. The awakening of human minds has always been, and will continue to be, an essential part of a more expansive cosmic awakening.

Mind is a definitional human trait, but modern archaeonomic materialism, by reducing mind to mindlessness, raises the question of whether and how we may rightly trust our minds to lead us to right understanding — that is, to truth. To justify trust in our minds, we have to renounce archaeonomic materialism; and if we want to connect our minds to the history of nature, we have to renounce analogical dualism. Our thought processes, if we are to justify our trust in them, cannot be reducible to mindless atoms or attributed to alien sparks of light that have drifted into the universe from above. Thought is fully natural. Yet nature itself is, from the start, an adventure of awakening. If we look at things from an anticipatory perspective, the whole universe—including our minds—is already in the grasp of a promise dawning on the horizon of the not-yet. It is because our minds, along with the whole universe, are being carried away by an indestructible rightness (infinite being, meaning, truth, goodness, and beauty) that we may justifiably, though always conditionally, put our trust in them.

NINE

Freedom

> I do not at all believe in human freedom in the philosophical
> sense. Everybody acts not only under external compulsion
> but also in accordance with inner necessity.
> —Albert Einstein

> Only if outward and inner freedom are constantly and consciously
> pursued is there a possibility of spiritual development and perfection
> and thus of improving man's outward and inner life.
> —Albert Einstein

IN DENYING THE EXISTENCE of a personal God, Einstein rejected the biblical belief that the world is created by divine power and cared for by divine love. While excluding—in the name of science—consideration of the existence of a personal God and the possibility of divine action in nature, he refused also for scientific reasons to accept the commonsensical belief in human freedom, or what he sometimes called "free will." How, he wondered, could people possibly have the faculty of free will, if the universe—of which we are a part—is run by unbending physical regulations? Addressing the Spinoza Society of America in 1932, Einstein declared that "our actions should be based on the ever present awareness that human beings and their thinking, feeling, and acting are not free but are just as causally bound as the stars in their motion."[1]

We humans, because of our powers of thought, are exceptional in the animal kingdom, but we are not exempt from the laws of nature. Physical lawfulness, Einstein assumed, covers everything, and humans are locked into a closed spacetime system with no way out.[2] What we take to be free will, therefore, is an illusion no less woolly than our impressions of the irreversible passage of time and of a God who answers prayers. Countless contemporary scientists and philosophers agree.

The word "freedom" has at least three meanings, all of which Einstein found unacceptable. Freedom may mean something we have, something we are, or something that has us. In the first sense, freedom is a human faculty that allows us to decide on one course of action rather than another. We call this faculty freedom of choice. In its second sense, freedom is not a mere faculty but a definitional human trait, one that sets us apart from other living beings, whose behavior is guided solely by instincts.[3] In this sense, freedom is a special mark of human dignity. In the third sense, freedom is the infinite source and destiny of finite being. Freedom, in this sense, is another word for God, who creates not out of necessity but out of unforced love.

In its third, theological, sense, freedom is so unrestrained, so devoid of self-preoccupation, that it spontaneously makes room for something other than itself to exist. Creation is a consequence of the divine kenosis (self-emptying love). It is out of God's infinite freedom, Christians believe, that the universe came into being. Infinite freedom does not compel the universe to exist, nor does it force creation to take on a predetermined shape. Rather, out of superabundant generosity, it calls the world into being and gives creation the opportunity to contribute to realizing its own identity, autonomy, and destiny. Infinite freedom gives itself away to the created world without reservation and still remains itself, for its essence is to love without limit. It is in response to the horizon of infinite freedom that the universe has become a drama of awakening. Freedom in this third sense is not something we have but something that has us.

Einstein denies the existence of freedom in all three senses.

Freedom of choice. For Einstein, the inviolability of nature's laws is proof that free will, freedom of choice, cannot be real. He was never able to reconcile freedom with relativity. Einstein scholar Matthew Stanley explains why:

> According to general relativity, all the objects in the universe, from cookies to badgers to every atom in your body, have a path through space-time called a worldline. Your worldline bounces

from event to event—intersecting with the worldline of the coffee shop before intersecting with the worldline of your boss before intersecting with the worldline of your ride home before intersecting with the worldline of your bed. You ride your worldline from event to event, encountering each like a train coming into a station. We humans, limited three-dimensional creatures as we are, only experience those events one by one. But the full four-dimensional fabric of space-time doesn't—it "sees" all those events at once. A being who could perceive the true nature of space-time would see their future and past stretched out along their worldline. Past, present, and future would be only relative terms.[4]

In other words, as Stanley goes on to say, "general relativity presents us with a universe that is sometimes called *deterministic*. That is, the future is already set. We only see one little part of our worldline, so we think the future is not set." But an imaginary observer of worldlines would see that "the future already exists." If this is true, our sense of freedom has no basis in objective reality. "Surely," as common sense coaxes, "I can decide what I will have for breakfast in the morning, and therefore change the trajectory of my worldline from intersecting with a doughnut to intersecting with oatmeal? Relativity says no. There is no free will. This sense that you can alter your future is an illusion, one caused by our incorrect perceptions."[5]

Freedom as the ground of human dignity. To the extent that Einstein embraced determinism, he forbade attributing to humans the quality of freedom that traditional theology and philosophy made central to our unique sense of dignity and self-worth. It did not seem to bother Einstein that his determinism also logically undermined the foundations of democracy, according to which every human person is endowed with freedom and responsibility, qualities that give us humans a special significance among terrestrial beings. But Einstein did not consider freedom to be part of the definition of human beings.

At times, nonetheless, he seemed to assume that we are free after all. He insisted, for example, on the importance of cultivating a sense of moral rightness, implying that we have a choice not to do so if we prefer.

In lectures especially during and after the Second World War he urged his audiences to resist militarism and totalitarianism. It is essential to human dignity that we pursue "outward and inner freedom," he said. And yet he told us just as often that belief in freedom is no more realistic than belief in a personal God.

Einstein was torn between determinism and freedom, as the clashing quotations at the head of this chapter indicate. On the one hand, he insisted that we humans are part of nature, subject to physical necessity. On the other hand, by exhorting us to embrace social causes and moral values such as pacifism, he implied that we are free to choose. As far as I have been able to tell, the great physicist never endeavored to resolve the contradiction. He insisted that our sense of inner freedom is a "subjective" illusion. But why would he encourage his fellow mortals to exercise their freedom and responsibility if these faculties have no basis in objective reality?

Freedom as something that has us. In the third place, and most importantly, Einstein also denied the infinite freedom that classical theology has identified with God. One of the reasons for his denial of a personal God was that the usual definition of "person" includes the attribute of freedom. Whenever he spoke of God, Einstein meant the rigorous lawfulness that gives predictability to nature. Nature's lawful necessity makes the world comprehensible to science but at the price of ruling out the existence of infinite freedom. Einstein never fully renounced his early attraction to Spinoza's belief in the eternity and necessity of the universe. Furthermore, in speaking of God as "not playing dice" with the universe, he was saying, in effect, that the universe is grounded not in liberating love but in eternal necessity. He was certain that nature's guiding principles are inviolable laws that permit no exceptions. Consequently, if the physical universe as a whole is tied so tightly to necessity, then every part of that universe, including the sphere of human existence, is similarly devoid of freedom.

Resolving the Contradiction

I believe the main reason Einstein could not reconcile science with freedom was that, like other modern determinists, he failed to distinguish the narrative aspect of the universe from its geometric substructure. Like other determinists, he assumed that the laws of nature are physical forces driving material bodies, including human beings, toward predetermined outcomes. Officially he left no room in nature for indeterminacy of any kind or degree. His determinism was deeply entrenched in modern scientific culture—specifically, in archaeonomic habits of thought.

There is, however, no scientific or logical basis for determinism. The belief stems from an unjustifiable modern attraction to mechanistic metaphors for nature. Mechanism, I am arguing, can be abandoned as soon as we learn to understand the cosmos as a drama of awakening. Einstein never looked for dramatic intelligibility in the universe but remained content with pure geometry and the postulate of physical determinism. Suppose, however, that the universe is not a machine but a drama. This metaphorical shift allows us to think of nature's inviolable regulations not as laws but as something like grammatical constraints. If so, the whole network of nature's invariant physical regulations functions comparably to a grammatical loom or grid into which a cosmic drama is continually being woven. The grammatical grid is invariant and inviolable, but the narrative or drama is indeterminate and its meaning unpredictable.

Grammar, after all, does not determine meaning. The rules of grammar are essential to expressing meaning in speech and writing, but they do not dictate the content of what is being said or written. In writing this book, I have to obey a set of grammatical rules as I move from one sentence to the next. I am not allowed to violate any of these constraints if I expect the book's content to be comprehensible. Nevertheless, the inviolable rigor of the grammatical rules does not keep me from weaving a different meaning into each sentence and chapter. As I move from one comment to the next, the rigid grammatical loom remains the same while the meaning changes freely.

Einstein, as we have seen, celebrated the universe's comprehensibility—its capacity to carry a meaning—and he rightly sought to express this comprehensibility by way of an ingenious geometry. But geometry is more analogous to the grammar of the cosmic drama to which we belong than it is to the content of the story. Grammar does not determine meaning, nor, analogously, do the invariant rules of nature determine the meaning or content of a dramatic universe.[6] If readers of this book focused only on the grammatical rules at work in my thinking and writing, they would miss any meaning the book may be carrying. Analogously, whenever scientifically educated experts focus only on the geometry of relativity, as in Stanley's example of the bouncing worldline, they fail to notice the dramatic cosmic story that is entwining itself in the geometry of spacetime.

By concentrating only on the grammar of relativity, determinists ignore the interior striving of living organisms that makes the universe dramatic. They even fail to notice the poignant drama of a great mind freely striving to make geometric sense of the cosmos. They do not seem to notice that Einstein's struggle to make the cosmos mathematically intelligible was as much a part of the cosmic drama as the birth of stars and the evolution of life. Contemplating the invariance of nature's laws and the splendor of relativity's geometry cannot tell us concretely what the universe is or what it is all about. The striving, the momentary successes and failures such as those that make Einstein's personal life so dramatically interesting to his fellow humans—these all slip through the wide meshes of his field equations. So also does the great drama of cosmic awakening of which the emergence of human freedom is a vital aspect.

Freedom, Cosmos, and Creed

Christianity assumes that our universe is the free creation of God and that freedom is an essential mark of human existence, a result of our being created in "the image and likeness of God" (Genesis 1:27). It claims uncompromisingly that we humans are responsible moral beings at liberty to make our own choices. Since we are free to be bad and not just good, we are in need of forgiveness—which also has to be given freely. Christian teachings on forgiveness allow that human freedom is limited and diminished by sin, age, sickness, and other circumstances. But a right appreciation of human dignity requires the affirmation of our core freedom and moral responsibility in the presence of others and God. So we are left with the question of how theology after Einstein may plausibly affirm the reality of freedom in all three senses—as choice, as an intrinsic part of our self-identity, and as a liberation of the whole of nature—especially if we agree with science that our existence is part of the physical universe. If we belong to nature, and if the laws of nature are inviolable, can we honestly defend the idea of human freedom?

I believe we can. The existence of freedom is compatible with contemporary science if the claims I have been making in this and previous chapters are reliable: first, that the universe is an open-ended story rather than a fixed and frozen state of being; second, that our specifically human existence is fully part of an ongoing cosmic awakening; third, that life, mind, and freedom are open to an anticipatory understanding rather than a purely analogical or exclusively archaeonomic reading; fourth, that the laws of nature are not deterministic physical forces but immutable regulations analogous to grammatical constraints; and, fifth, that the universe's true identity is to be found not in its physical past or in its geometric design but in the not-yet of its dramatic becoming. It is because we humans are conditioned not only by invariant physical rules but also by the universe's dramatic openness to the indeterminate horizon of the not-yet that we are given the gift of freedom.

An anticipatory understanding of nature makes room for freedom in a way that is completely compatible with science, though not with archaeonomic materialism. The archaeonomic stance leaves no room for freedom, since it reduces every present and future state of nature to an initial cloud of mindless elements governed by unbending laws of nature. Natural phenomena, including humans, are, in this reading, so fully enmeshed in physically fixed worldlines that no openings exist anywhere for freedom. Analogy, our second reading, allows for freedom but only by severing human existence—in some unspecified way—from the physical universe in which we presently abide. Anticipation, for its part, allows for the existence of finite human freedom in the first and second senses of the term because it holds that the universe is an awakening to infinite freedom in our third sense. In an anticipatory universe the emergence of human freedom is not a tear in the tissue of invariant physical habits, nor a side effect of quantum indeterminacy, nor a supernatural interruption of the continuum of physical causes. Rather, freedom is an attribute of a universe that is now in the process of being released from the fixity of its past by an awakening to what is not-yet. Human freedom exists to the extent that each of us participates in the drama of cosmic awakening.

In fleshing out this proposal, I argue that the deterministic freedom-bashing beliefs associated with archaeonomic materialism are both logically mistaken and scientifically shallow. It is not freedom but determinism that turns out to be the stultifying illusion. Next, I argue that contemporary analogical and dualistic attempts to rescue human freedom from the modern materialist debunkers are morally respectable but intellectually unsuccessful. Analogy's reactionary Platonic reading of the cosmos may be worthy of our attention, but it fails to take theological advantage of the new narrative understanding of the universe given to us by the natural sciences. Finally, I argue that the anticipatory/Abrahamic reading of our unfinished universe allows us fully to embrace science, including evolutionary biology and Big Bang physics, without having to sacrifice the biblically based doctrine of human freedom and the reality of time. Overall, I contend that our trust in the reality of freedom is given new life and wider scope than ever if we locate it within the dramatic texture of an awakening universe. To back up this anticipatory position, I will be comparing it with the archaeonomic and analogical perspectives on human freedom.

Archaeonomy. Not science but the archaeonomic stance of modern thought contradicts the idea of freedom in all three senses. Today the archaeonomic reading of natural history takes for granted that our decisions and behavior, including what seem to be morally motivated acts of generosity, are not our own. Whatever happens in the universe, we are instructed, is due to physical compulsion beyond our subjective control. And subjectivity, according to archaeonomy, does not even have real existence.

Einstein, following both Spinoza and archaeonomic interpretations of nature, was a determinist from his youth on, and he was convinced that determinism is reinforced by the spacetime geometry of general relativity. As we have seen, there is also an atomistic twist to Einstein's natural philosophy: the idea that nature at bottom is nothing more than a continual reshuffling of primordial physical elements moving in blind obedience to invariant laws. This academically popular perspective persuaded Einstein that scientifically educated people must give up once and for all both the Platonic and the Abrahamic notions of freedom.

The archaeonomic denial of freedom seems as entrenched as ever today, especially in fields related to neuroscience. In a recent book the materialist philosopher and neuroscience expert Patricia Churchland, for example, affirms anew what she takes to be the necessary connection between determinism and contemporary science.[7] Scientific method, she claims, requires that a neuroscientist embrace the doctrine of determinism simply because the human brain and its mental functioning are webbed fully into the deterministic material makeup of nature. It follows that because of the invariant laws of nature, whatever "decisions" individuals make cannot really be free and that moral responsibility is nonexistent.

To the devout determinist, freedom, like God, is an illusion that we humans have invented to support a groundless sense of self-importance. The ideas of freedom, God, and human dignity are pure fictions that may give us the impression that we are special. In the past, these fictions may have been adaptive in an evolutionary sense. Now, however, science has demonstrated that they are fanciful. To a Darwinian, the conventional belief in freedom is an adaptive illusion, but now we may learn to live without it. Science, the only reliable arbiter of truth, has become the best means we have of adapting to the world around us. If we become sufficiently informed scientifically, we shall automatically abandon the smug illusions of freedom, dignity, and God once and for all.[8]

Isn't it a good idea, nevertheless, to live and act as if we are free, even though science knows better? Several years ago, *The Atlantic* ran an article entitled "There's No Such Thing as Free Will," summarizing the belief by archaeonomic determinists that freedom of choice is an illusion. "But," as the article's subtitle instructs its readers, "we're better off believing in it anyway." The author, Stephen Cave, writes:

> In recent decades, research on the inner workings of the brain has helped to resolve the nature-nurture debate—and has dealt a further blow to the idea of free will. Brain scanners have

enabled us to peer inside a living person's skull, revealing intricate networks of neurons and allowing scientists to reach broad agreement that these networks are shaped by both genes and environment. But there is also agreement in the scientific community that the firing of neurons determines not just some or most but all of our thoughts, hopes, memories, and dreams.

Cave, however, is uneasy with this determinist perspective. "If moral responsibility depends on faith in our own agency," he asks, "then as belief in determinism spreads, will we become morally irresponsible? And if we increasingly see belief in free will as a delusion, what will happen to all those institutions that are based on it?"[9]

The same author cites recent surveys indicating that people will be more generous and happier if they believe they are acting freely than if they believe their actions are predetermined. Science, Cave admits, seems to have demonstrated that freedom does not exist, but in doing so it has raised disturbing moral questions. Following the philosopher Saul Smilansky, Cave asks whether we shouldn't "hide this sober truth from the masses of scientifically illiterate people." Smilansky warns that "there is something drastic, even terrible," about the idea that we are not free. Will people behave themselves, he asks, once they learn from science that their own sense of freedom is an illusion? He concludes that "if the choice is between the true and the good, then for the sake of society, the true must go." Accordingly, only a carefully selected body of the scientific elite—the "fully initiated"—should be allowed to look into the dark abyss of a deterministic universe.[10]

The archaeonomic debunking of freedom by experts in the neurosciences now seems to be supported by almost every other field of scientific endeavor, including especially biology. The suffocating sludge of determinism started gathering intellectual momentum in the field of physics almost four centuries ago, but it has now oozed out into the life sciences, psychology, and sociology. Archaeonomic determinism has affected scientific and philosophical thought all over our planet today. Its appeal lies mostly in its promise to make the world fully predictable, an expectation that underlies Einstein's own preference for a law-bound universe.

Einstein did not seem to notice, however, that his personal journey of scientific discovery was itself dramatic. The path that led to his discovery of relativity was an epic of ups and downs, fits and starts, successes and failures. Nor did it occur to him that his personal story participated in and contributed to the larger drama of an awakening universe. He found no place in the geometry of the universe for his own life-story. His exclusivist interest in geometric coherence left no place for freedom, since it left no place for the irreversible passage of time essential to the existence of freedom. Nor did Einstein ask what it means that the universe has given rise to beings like him who can be carried away at the core of their existence by the indestructible rightness of mystery, truth, beauty, and goodness.

Weaving itself onto the grammatical loom of Einstein's spacetime universe is the ongoing drama of lives—most of them riddled with struggle, suffering, success, and failure—awakening to the open horizon of the not-yet. It is in this never-ceasing, temporally irreversible, dramatic openness to the not-yet, I propose, that freedom finds its home in a cosmos still aborning.

Analogy. At least Einstein tried to honor the value of truth-telling. It would never have occurred to him to emulate the shocking condescension of contemporary scientific and philosophical determinists who—out of "compassion" for the ignorant masses—try to conceal from them the "truth" that freedom does not really exist and that it would be advantageous to everybody if the less elite intellectual classes of humans are not exposed to the harsh doctrine of determinism! Fyodor Dostoyevsky's Grand Inquisitor had the opposite concern: How can we keep the masses from realizing that they have freedom? Maybe one way is to keep supporting the doctrine of materialist determinism.

Contemporary devotees of analogy usually have no difficulty agreeing with Einstein that the spatiotemporal world is a closed causal continuum devoid of freedom. Some of them accept physical determinism as far as nature is concerned, but they allow for the reality of human freedom. They do so, however, only by locating the core of human existence in some other metaphysical space than that of the physical universe. Freedom, in the analogical interpretation, is a trait that hovers over physical existence instead of emerging from it. Freedom somehow accompanies our bodily existence. In classical versions of analogy, freedom resides in an immaterial "soul" that never fully belongs to the temporal physical world. The soul is said to be free precisely because it is not subject to the laws of physics or caught up in relativity's inflexible worldlines.[11]

Although the analogical stance appeals to many staunch defenders of freedom, it is ultimately unsatisfactory intellectually because it fails to connect human beings to nature in the seamless way that science after Darwin and Einstein has done. Analogy boasts that humans have freedom of choice, but it does so only at the price of ignoring the physical, biological, and neurological factors that underlie our mental activity, including our capacity to make moral decisions.

Anticipation. Contemporary science has demonstrated that we humans are products of deep time, fully part of nature rather than an interruption of it. Human persons belong completely to the universe. But if we are part of the universe, and the universe is bound by rigid laws, can we reasonably claim to be free and responsible beings?

We may find space for both science and freedom, I suggest, by locating ourselves on the brink of the not-yet to which an anticipatory universe is always awakening. Freedom, I submit, shows up in the dramatic rather than the grammatical fabric of the universe. Freedom is not an illusory feeling welling up causally from the fixed atomic or material past. Nor is it a mysterious cloud floating down from above. The place to look for freedom is in the openness to the future of a universe that is still awakening in real time. We should look for freedom not in spacetime geometry but in the dramatic awakening of the cosmos made possible by the irreversibility of time. Freedom is possible only if the cosmos can become new in each moment of its existence. Newness, however, shows up in the drama, not the grammar, of nature.

If you are writing a novel, your narrative must adhere to strict grammatical rules. But your scrupulous adherence to the rules of writing does not keep you from creating a meaningful literary product whose content no reader could have predicted in a million years. Analogously, the cosmic story runs along millennium after millennium without straying beyond the geometric constraints defined by relativity. Yet prior knowledge of relativity or any other scientific regulations would not be enough to let you predict, even in principle, precisely what will happen in the drama of cosmic awakening. Without breaking or bending any fundamental physical laws, the universe has from the beginning been open to unpredictable dramatic outcomes such as the striving of life, the awakening of minds, and the responsible exercise of freedom.

Freedom arrives not in the rules but in the universe's narrative openness to what is not-yet. Freedom is not an interruption of nature's laws—which is how Einstein caricatured it. Freedom attaches itself to the irreversible temporal flow that releases the universe each moment from the fixity of the past. Einstein's denial of freedom is of a piece with his love of eternal necessity and his denial of irreversible time. If time is real, however, the actualizing of freedom turns out to be a central motif in the cosmic drama. What is really going on in the universe is the gradual emergence of freedom—and all its dramatic side effects, including opportunities for both responsible action and monstrous evil.

Along with life, mind, and God, freedom is not something we can ever discover or locate archaeonomically. Just as archaeonomy is blind to the cosmic story, it is also blind to freedom. If you are looking back only in the direction of the universe's fixed past, freedom will never show up. Freedom is not an objective thing or property that pure scientific analysis can ever conceivably discover or rule out. Instead, freedom is a mode of being that is experienced in the interior life of conscious subjects. Attempts to account for the "inside" experience of freedom through the grammar of what is observable "outside" will inevitably fail.

Earlier we saw that something similar is true of our knowledge of life. By examining the molecular and cellular infrastructure of living beings we cannot expect to apprehend the striving that makes life dramatic. Nor can we encounter the reality of freedom by looking at the tissue in human brains and nervous systems. Like the striving that defines life, freedom will never show up on an archaeonomic map of nature. Ultimately it is because the whole universe is always opening dramatically to what is not-yet that freedom, in all its fragility, slips into the cosmic story. Through the medium of human faith and hope, as we shall see, the universe is further liberated from fixity to the past and can thus be a suitable habitat not only for life but also for growth in freedom.

Both archaeonomy and analogy fail to make a place for freedom in the natural world. By ignoring the passage of irreversible time, each in its own way gives us a universe that is already finished—either in eternity (in the case of analogy) or in the fixed past (in the case of archaeonomy). In either case, there is no dramatic openness, no room for the arrival of anything truly new in the passage of cosmic time. Both archaeonomy and analogy are time-suppressing worldviews, unfriendly to freedom because, first of all, they are unfriendly to the fact of irreversible time and its trending toward the horizon of the not-yet.

So the debate about freedom is inseparable from debates by scientists, philosophers, and theologians about the objective reality of time. If time is unreal, there is no room for the not-yet, the place where freedom breathes. By freezing the cosmos into the fixed tracks of its opening moments, archaeonomy allows nature to have no future outcomes that are truly distinct from what is already complete in the beginning. To protect against the anxiety of existing in perishable time in an unpredictable world, archaeonomists deny the real existence of both duration and freedom. The two denials go together, and they join up with the more fundamental denial of the reality of God.

Summary

Freedom can exist only in a universe that is permitted to become new. The universe can become truly new, however, only if time is an irreversible passage. Einstein tried to eliminate the passage of time by dint of geometry, but Big Bang cosmology has given time back to us abundantly. "Everything is determined," Einstein persisted, "by forces over which we have no control. . . . We all dance to a mysterious tune, intoned in the distance by an invisible Piper."[12] An anticipatory stance maintains, however, that the inviolability of nature's regulations does not mean that the cosmic story is fully predictable. Even if the laws of nature are unchangeable, their inflexibility does not rule out the emergence of dramatic new moments, chapters, and epochs as the story proceeds.

It is not surprising that determinists like Einstein cannot find anything like freedom in the geometry of nature. Freedom cannot by definition show up there. Determinists are transfixed by the grammatical loom that holds nature together, leading them to overlook the emergence of freedom as part of the greater drama of cosmic awakening—even as it is occurring locally in the activity of their own minds.

The laws of physics and chemistry do not have to be suspended in order for life, thought, and freedom to be objectively real developments in the drama of an awakening universe. Physically speaking, as Einstein rightly insisted, the laws of nature have been, and will continue to be, operating predictably and universally. But since the cosmos is a story and not a mechanism or a reshuffling of elementary particles, its inviolable laws function more like rules of grammar than deterministic forces. Their predictable invariance is completely compatible with the dramatic arrival and survival of novelty and freedom. Novelty and freedom come into the universe not as part of its ageless grammar but at dramatic intervals in the cosmic story when—carried along by irreversible time—the already is greeted by the not-yet.

TEN

Faith

To the [sphere of religion] belongs the faith in the possibility that the regulations valid for the world of existence are rational, that is, comprehensible to reason. I cannot conceive of a genuine scientist without that profound faith.
—Albert Einstein

Science without religion is lame, religion without science is blind.
—Albert Einstein

... there is yet faith / But the faith and the hope and the love are all in the waiting.
—T. S. Eliot

NOTHING STANDS OUT MORE emphatically in Jesus's teachings than his exhortation to have faith. In the New Testament to have faith (*pistis*) means "to trust in" or "be loyal to" something or someone.[1] The Nicene Creed begins with the words "I believe." This is a translation of the Latin *credo,* whose roots are *cor* and *do:* "I give my heart." God—the superabundant source of extravagant promises—is the one to whom we may give our loyalty and our hearts without reservation.

In Christian religion faith is not, in the first place, a list of beliefs but a heartfelt surrender to what is indestructibly right. We may explore the meaning of faith through psychology, sociology, anthropology, biology, history, classical theology, and other disciplines. After Einstein, however, we may at last seek out the meaning of faith also by way of cosmology. Our scientific awareness that the universe is still coming into being allows us now to think of the phenomenon of faith as marking yet another new emergent epoch in an awakening universe. The universe, too, is justified by faith.

Einstein had no use for a personal God, or freedom, or the irreversible flow of time, but he did make room for faith. For him faith meant putting trust, not in a personal God, but at least in something of imperishable importance. Sometimes he even called faith "religion," but religion did not mean loyalty to an interested deity. Scientists, Einstein advised, may cultivate faith if they understand it as a cosmic religious feeling.[2] In this sense, faith for Einstein meant reverence for the phenomenal universe, which he took to be the outward manifestation of something timelessly intelligible beneath appearances. Moreover, when Einstein spoke about the necessity of religion or faith, as, for example, when he said, "Science without religion is lame, and religion without science is blind," he was implying that doing science rightly requires on the part of every inquirer a devout appreciation of the (mathematical) harmony that pervades the cosmos. In an interview in 1930, Einstein declared, "I am of the opinion that all the finer speculations in the realm of science spring from a deep religious feeling, and that without such feeling they would not be fruitful."[3]

It is understandable that scientists, while actively engaged in their work, may be oblivious to the sentiment of faith or deep religious feeling supporting their steady empirical resolve. Reflective scientists, however, may at times notice how their spontaneous personal trust in the importance of right understanding links them at least vaguely to the larger and older human communities of faith. Honesty should lead scientists to acknowledge, with Einstein, that only an unflagging faith in the universe's elusive intelligibility and a surrender to the value of truth telling can keep them on course in their various pursuits. It would be no insult to the nobility of their calling, then, if scientists were to own consciously and gratefully the high degree of faith that energizes both their lives and their work.

Einstein thought that in striving for insight and truth, scientists have an implicit loyalty to (or faith in) indestructible rightness. A scientist's fidelity to the value of truth, for example, is usually only tacit, but without a commitment to truth, science is impossible. Few scientists would die for truth, but many would risk their careers for it. Their commitment to truth is not unlike the trust religious people throughout the ages have had in something indestructible. Unlike Einstein, most scientists are inattentive to the faith, or "religious" trust, underlying their quest for understanding and truth. They seldom pay explicit attention to the rich web of beliefs woven into their inquiries. Judging, however, from what Einstein has written about religion and science, I think he would have agreed that there are at least four kinds of faith commitment quietly at work in all fruitful research:

1. Faith that the universe is intelligible.
2. Faith that truth is worth seeking for its own sake.
3. Faith that honesty, humility, generosity, and openness in sharing one's ideas and experiments are unconditionally right.
4. Faith that one's own mind has the capacity to grasp intelligibility and distinguish what is true from what is false.

Einstein confessed that he had found "no better expression than 'religious'" for the confidence he had in "the rational nature of reality insofar as it is accessible to human reason." The fourfold faith, he would agree, is absolutely essential to doing science rightly.

I want to propose further that faith is also an essential part of the drama of cosmic awakening. Faith, cosmically speaking, has its roots in the striving of life, and it underlies the right functioning of human minds. In this sense, faith means loyalty to the ideal of rightness. Faith is also the willingness to wait for the arrival of deeper truth, greater goodness, and wider beauty. Faith serves hope by giving it the patience to wait. Without the patience of faith our minds would be at the mercy of instinct, and hope would decay into wishing. Faith serves the drama of cosmic awakening by protecting the mystery of indestructible rightness from being replaced by impatient, premature illusions of certitude. Instead of being opposed to truth, faith keeps thought from drifting off to sleep.

The coming of life started transforming the universe into a drama of striving 3.7 billion years ago, introducing into it the possibility of both success and failure. Later on, the coming of thought turned the universe into a drama of awakening to meaning and truth. Almost simultaneously with the emergence of thought, the phenomenon of faith introduced into the universe a willingness to wait courageously for the coming of the not-yet. More than anything else, faith means waiting patiently for future possibilities to be realized.

Three Kinds of Faith

Faith is faintly present, however, even when thought is unwilling to wait.

Archaeonomy. Archaeonomy is itself a limited version of faith. It looks for a principle of coherence that will make everything intelligible. The faith that launches the archaeonomic adventure takes for granted that the ultimate principle of coherence lies in the remote cosmic past. Intelligibility resides in the beginning rather than in an eternal present or a still-uncertain future. Archaeonomy wagers that after a swift journey into the remote cosmic past by way of reductive scientific analysis, the quest for coherence will come to rest in a principle of unity and mathematical simplicity that ties everything in the universe together physically from the start.

Religious faith, archaeonomists insist, is unwarranted and perhaps even cowardly for believing that the universe has a meaning unavailable to science. Archaeonomists consider Christianity especially naive for waiting patiently for a future coherence.[4] To entrenched scientific materialists, the world's future will finally amount to nothing more than what was already predestined by impersonal physical laws from the start. And if the future is already set, what would be the point of waiting? Waiting, at least to the archaeonomists, would not mean a soul-expanding vigil for the coming of something new and unpredictable. Instead, it would mean counting time to find out what has always been inevitable. No less impatient than strict pantheism, archaeonomic pessimism fails to allow that time carries with it emergent outcomes that were not implicit in the cosmic past.[5]

Archaeonomists claim, with a remarkable sense of certitude, that religious faith is a mark of human weakness, an illusory product of cerebral agitation and unacknowledged evolutionary adaptation. At the same time, they exhibit an unquestioning faith in the authority of their own minds to make this claim. Is faith in their cognitive faculties justifiable? As I argued earlier, I believe it is, but not by virtue of their archaeonomic metaphysics. They have forgotten that archaeonomy had already reduced all minds, including their own, to the elemental mindlessness from which everything in the universe is supposed to have sprung. In the very act of claiming that scientific materialism is true and religion false, they rely on faith in the power of their own minds to find truth and expose illusions. They are not wrong to have such faith in their minds. Their archaeonomic worldview, however, does not have the logical consistency or explanatory amplitude to justify that faith.[6]

Analogy. Analogical faith, by contrast, claims that our minds and souls are somehow already grasped by the indestructible rightness that exists beyond the physical universe and the perishable world of time. We can have faith in our power of thought because our minds exist not only in the physical universe but also in eternity. The more fully our minds remain immersed in worldly time, the less trustworthy they are. Only if we undergo moral and religious conversion to the supernatural can we trust that our minds will make contact with truth. So the life of the mind is not independent of the practice of virtue and faith. A moral and religious choice is essential to letting our minds be taken captive by imperishable meaning, goodness, beauty, and truth. In this sense, faith in God is essential to our arriving at right understanding, right action, and right living.

Einstein comes remarkably close to embracing an analogical faith at times. He professes to have a "firm belief, a belief bound up with deep feeling, in a superior mind that reveals itself in the world of experience." This he says "represents my conception of God. In common parlance this may be described as 'pantheistic' (Spinoza)."[7] Like analogical theology, Einstein admits the limitations of our senses and minds, and he celebrates the world's intelligibility, even though it never lies fully within our grasp. His religiousness consists of "a humble admiration of the infinitely superior spirit that reveals itself in the little that we can comprehend of the knowable world."[8] He even sometimes calls this spirit God. But he is talking about Spinoza's God—whom the pantheist identifies with nature—and not the God of Israel, "who concerns himself with the fate and the doings of mankind."[9]

Analogical faith grounds the world's being and intelligibility in the radical transcendence and personhood of God. God is not the world, and the world is not God, but analogy contends that only formal belief in God, in Truth-Itself, can fully justify the faith we have in the capacity of our minds to discover meaning and truth. Even though the mind of God has endowed the universe with comprehensibility, we humans, as long as we exist in time, may expect to have only a limited understanding of it. To analogy, therefore, it is the height of impiety, not to mention an impossible epistemological burden, for the finite human mind ever to aspire to complete scientific mastery of the totality of being. Here again, Einstein would agree. His love of mystery is too reverential to let him believe that science could ever do more than scratch the surface of nature's inexhaustible intelligibility.

Einstein accepts analogical theology's claim that the universe is comprehensible but not its faith in an infinite, transcendent, personal intelligence that freely endows creation with comprehensibility. He differs from analogical theology in his insistence that the ultimate principle of intelligibility is purely geometric and impersonal. Einstein would agree with analogical theology, however, that it is not up to us humans to give meaning to things, contrary to what modern thought has often assumed. And he would agree with classical theology that faith is the state of allowing our minds to surrender humbly and joyfully to a meaning and truth that are mostly hidden. In tune with analogical religious faith, he wants to avoid getting tied down by what takes place in time. He seeks to be elevated by faith above the passage of time into the sphere of timelessness where full intelligibility lies. Geometry alone is sufficiently equipped to carry out the directives of his analogical faith.

Anticipation. An anticipatory faith believes that our minds cannot arrive at right understanding solely by coming to rest in the simplicity of the cosmic past or by looking away from time by way of geometry. Instead, anticipatory faith links itself to the universe's long dramatic journey already under way. It is aware that the universe's comprehensibility is a goal for which we always have to wait. Faith lies mainly in the waiting. Since the world is not yet fully intelligible, anticipatory faith is aware that final coherence is still in the process of formation. Patience, not obsession with certitude, is the key to reconciling our minds with the passage of deep time.

Anticipatory faith refuses to escape from time, whether by resting in the atomic simplicity of cosmic beginnings or by escaping into eternity.

Rather, anticipation takes the risk of journeying along with the universe, in search of a narrative coherence now hidden in what is not-yet. Anticipatory faith does not look for an encompassing principle of coherence by going back in time to the subatomic simplicity from which the universe began to awaken billions of years ago. Nor does it look for preexisting intelligibility in a timeless heaven of ideal forms already perfected. Since the universe is not yet complete, and its forms are always in flux, anticipation waits instead for a dramatic meaning that for now may seem to be only vaguely taking shape up ahead.

Patient expectation, I cannot emphasize enough, is epistemologically essential to the anticipatory understanding of everything. Anticipatory patience assumes that only the most trivial truths land in our laps without waiting. The deepest truths always take time and come with a price. What interests an anticipatory faith, then, is not whether the universe is comprehensible but when. Accordingly, anticipation asks Christians to recite their Creed in the spirit of expectation and not just contemplation. This is because the realm of "all things, visible and invisible," has never yet been fully established. Creation is still happening, and the divine communion with time is ongoing. The universe's full intelligibility needs to be approached and protected for now not only by faith but also by the virtues of humility, temperance, and hope.

Faith in One God

All three readings of the universe are expressions of our human longing for coherence, but only anticipation has the patience to wait for the requisite unity to become actual in the universe. Anticipatory faith, in embryo, arose as soon as humans felt vaguely an invitation to be led beyond present and past understanding to a meaning not yet fully realized. By way of symbols, rituals, and myths our ancestral religions, almost from the start, sought to bring narrative unity to the scattered moments and episodes of people's lives. It was the function of myths and other kinds of narrative to weave the otherwise dispersed episodes of human experience into an overarching coherence.

Eventually, somewhere between two and three thousand years ago, out of the long and tangled web of mythic narratives there emerged the story of faith in the oneness of God. The Creed's affirmation—"I believe in one God"—expresses Christianity's confidence that all things somehow hang together. But, again, the question is when. Each of our three readings gives a different answer. Archaeonomy, as the word suggests, trusts that cosmic coherence, was established in the beginning (arche) and can be discovered only by our going all the way back in cosmic time. Analogy trusts that the principle of the world's intelligibility lies beyond time altogether, having already unified all things eternally, although in a way that is now mostly inaccessible to finite minds immersed in matter and time. An anticipatory faith, however, trusts that cosmic coherence lies in the direction of what is not-yet in the passage of time.

To encounter the deepest meaning of things, an anticipatory faith requires our remaining in time, in a state of vigilant expectation. Belief in "one God," as professed in the Creed's first article, therefore, is a matter of patient expectation. Faith in God entails patient, but not passive, waiting for a comprehensibility not yet fully actualized. Abrahamic monotheism is not a search for certitude but an expression of the human need to wait in hope for the final dramatic unifying of all things. The anticipated coherence is not an initial physical simplicity, as archaeonomy projects, but a weaving together of all previous events into a complex tapestry that preserves and enhances the individuality and difference of every creature that emerges in time.

Anticipatory faith gives dramatic meaning to our lives even here and now by encouraging us to keep our attention fixed not on the past or the eternal present but on what is yet to come. It is this Abrahamic brand of expectation that animated the earliest Christian communities and that may also lift up our own hearts and minds. Unfortunately, the early Christian spirit of anticipation has often given way to a world-fleeing nostalgia for timelessness in both religious and secular cultures.[10]

While Einstein had faith in the universe's comprehensibility, he had no use for monotheistic theology. Why not? His hesitation cannot be justified by the claim that theology is based on faith, and science is not. Both science and theology require faith in the world's intelligibility without which they could not even get off the ground. Nor can Einstein's antitheism be justified by the claim that theology's personal God appears to act by interrupting the timeless laws of nature. After all, as I have observed, theology does not require the violation of habitual physical routines in order for dramatic things to happen in the universe. Rather, the reason for Einstein's distaste for monotheistic faith lies mainly in his belief that the universe's intelligibility can be packaged well enough in a mathematical format. Once again, this bias is fortified by a prior denial of the irreversibility of time.

With Galileo and Descartes, during the first half of the seventeenth century, a neo-Pythagorean quest for understanding things of nature exclusively in the mode of geometric coherence had already begun to edge out what remained of the anticipatory Abrahamic faith. Why was it so easy for this transition from a dramatic to a geometric ideal of understanding to occur at the beginning of the modern age?

I believe it is mostly because Christian theology and philosophy had never fully accepted the reality of time. The modern flight from time into geometry is mostly due to habits of mind sculpted by long exposure to the analogical stance in classical Christian interpretations of the world. For centuries the analogical reading had persuaded philosophical and religious minds to locate the source of the world's intelligibility in a place untouched by time. Having hidden the passing of time for centuries under the shadow of an eternal divine present, the analogical vision eased the way for modern scientists to ensconce the physical universe entirely inside the timeless sphere of pure geometry.

In spite of ongoing attempts by archaeonomic scientists and philosophers to deny the narrativity of matter, nature's dramatic constitution is now becoming increasingly evident. By unwittingly uncovering the dramatic texture of nature, Einstein's cosmology lets us understand faith, including the tacit faith of scientists, as a dramatic new development in an awakening universe. In the course of cosmic history, life long ago emerged out of lifeless matter, and thought more recently arose out of life. But simultaneously with thought, faith came into time as an essential part of the cosmic story. Through the exploratory adventures, and sometimes misadventures, of faith the whole universe has long been seeking a dramatic communion with indestructible rightness.

After our religious traditions became literate over the past several thousand years their teachers sometimes thought of the cosmos as a book or scroll. The scroll of nature, they assumed, is something to be read, and if ordinary texts can be read at different levels of depth, so, too, can the passage of nature. But to find out what our holy books mean, minds and hearts had to change. And until modern times religions assumed that to find out what's really going on in the universe, an ongoing transformation of hearts and minds was essential too.

What meaning does the book of nature carry deep down? If a classic literary text can be read at different levels of understanding, might this not be true also of our new thirty-volume scientific cosmic story? If so, would our own minds and hearts not be obliged to undergo an extraordinary transformation to put us in touch with the deepest meaning of the cosmic story? Refusal to allow for such a transformation—that is, to allow for faith—could leave us blind to the universe's real meaning.

Ordinary consciousness, because of its impatience, cannot peer into the depths of the universe—or the Scriptures. Nor can science. Archaeonomists, however, claim to have discovered what the universe is really all about without having to wait. Science, they believe, is now laying bare the universe's invariant physical regulations, pervasive geometric principles, and subatomic constituents for all to see. For the archaeonomists that is enough. There is no need to wait for anything else.

Anticipation, however, insists that full intelligibility resides not in origins but in what is yet to come, and for that we do have to wait. Anticipatory faith also gives us a new answer to the Kantian question of what we should be doing with our lives in the meantime. It is of great interest morally speaking that science has now demonstrated that we live in a universe that is still awakening and is therefore open to new possibilities of being. Each of us can find moral significance in our lives by contributing to this cosmic awakening locally. Kant's universe functioned as a relatively featureless backdrop to human life. For him the moral life was analogically grounded in a timeless divine imperative awakened in each person's heart from up above. An anticipatory faith, by contrast, looks at the universe, not the human mind or heart, as the primary subject of awakening. Consequently, the content of moral life is not just that of improving personal existence but, also more broadly, of contributing to the ongoing transformation of the universe.[11]

Summary

Nature, as we look at it from an Abrahamic point of view, has been seeded from the outset with the promise of becoming *more*. It is not geometric design but the promise of fuller being that allows the cosmos to exist in the shape of an unfinished drama rather than as an architectural fait accompli. Since the story is still going on, we need not approach it with either archaeonomic pessimism or analogical impatience, but we can instead approach it with patient open-ended anticipatory expectation.

Einstein's thought blended archaeonomic with analogical assumptions, and he was fascinated by the fact that the universe is intelligible at all. For him, nature's meaning was reducible to the kind of comprehensibility that makes itself known only when we approach it with the simplifying and unifying power of geometry. "We may in fact regard [geometry] as the most ancient branch of physics," Einstein notes, confessing that without it he "would have been unable to formulate the theory of relativity."[12]

Yet Einstein's own theory of general relativity opened up the possibility that the universe, understood as existing in real time, has a narrative intelligibility that geometry cannot fathom. The universe is a story, and since stories are made to carry meaning, an intelligibility deeper than geometry may be fermenting beneath the surface of nature during the irreversible passage of time.[13] It is the function of faith to keep our minds open not only to geometric but also to dramatic meaning.

ELEVEN

Hope

> Is this, then, all that life amounts to—to stumble, almost by mistake, into a universe which was clearly not designed for life, and which, to all appearances, is either totally indifferent or definitely hostile to it, to stay clinging on to a fragment of a grain of sand until we are frozen off, to strut on our tiny stage with the knowledge that our aspirations are all but doomed to final frustration, and that our achievements must perish with our race, leaving the universe as though we had never been?
> —James Jeans

> Our world contains within itself a mysterious promise of the future.
> —Pierre Teilhard de Chardin

> I look forward to the resurrection of the dead and the life of the world to come.
> —Nicene Creed

NOT LONG AFTER THE DEATH OF Jesus, a new flame of hope kindled by belief in his bodily resurrection began to spread across the ancient Mediterranean world. The Christian hope for the "resurrection of the dead" brought a new appreciation of real time into world history. Still clinging to the ancient anticipatory stance of Abraham, the earliest Christians understood time as real, irreversible, and filled with meaning. They looked for the eventual return of Jesus at the end of time to bring all events to a glorious fulfillment. They hoped the ending would be soon, but they were willing to wait.

If the gradual rise of life, mind, faith, and freedom allows us to understand the universe as a drama of awakening, all the more so does the hope that springs eternal in human hearts. Christian resurrection-hope, as distinct from archaeonomic pessimism and analogical optimism, implies that the creation carries with it a unifying principle of meaning not yet fully actualized. Throughout much of Christian history, however, the analogical longing for timelessness has weakened Christianity's hope that the whole universe will have an eventual fulfillment. Given the fact of perishing, the allure of heavenly timelessness has generally been more attractive than the prospect of living fully in time in the expectation that creation will be healed and made whole. Even today, remnants of Platonic otherworldly optimism overshadow the strains of Abrahamic hope left pulsing here and there in the religious world. Meanwhile, the figure of Democritus continues to loom over intellectual life, giving us a world whose gates we may enter only after abandoning hope.

All this pessimism notwithstanding, I believe that questions about the meaning of time and reasons for hope are still stirring beneath the surface. At the end of the eighteenth century the philosopher Immanuel Kant was still asking the eternal questions: What can I know? What ought I to do? What may I hope for? These questions are still alive. How to answer the third of these, however, is more complicated now than in Kant's lifetime. After Einstein, it is hard to separate our personal anxieties about time and perishing from the contemporary concern among cosmologists about the eventual death of the universe itself. We realize nowadays, more certainly than ever before, that human history, the story of life, and the whole cosmic journey are a package deal. The hope that swells in human hearts is an ache that rises up from the heart of matter itself.

It is hard to imagine, though, how hope for the material universe can be reasonable and right if, as many astrophysicists surmise, absolute death is the final destiny of the whole story of nature. The universe, science tells us, is going to collapse owing to energy depletion in the far distant future. How are we to digest this bitter news? The analogical response is that we can always leap overboard into the sea of timelessness as the cosmic boat approaches final catastrophe. More often than not in Christian history, hope has meant the soul's longing to be transferred from the tortures of time to the bliss of eternity. As long as the soul can find salvation in an immaterial heaven after death, is this not enough? What happens in time or world history never lasts anyway. So what difference does it make whether or not the universe endures? Why is the long passage of time worth worrying about?

Not only analogical faith and theology but science, too, has had a troubled relationship with time. Theoretical physicists today, according to one of them, are generally more impressed by pure timelessness than by the concrete passage of time.[1] Even scientists who do not believe in God, or the soul, or in life after death, still sometimes enjoy turning their minds loose in the timeless theater of pure geometry. In this immaterial realm they find relief from the merciless momentum of irreversible time. It is satisfying enough for them to assume with Einstein that geometry rather than providence decides what the universe is all about.[2] "It fortifies my soul to know that, though I perish, truth is so," exclaimed the nineteenth-century poet Arthur Hugh Clough. A similar sentiment must uplift the hearts of many pure scientists. In this respect, a thinly disguised version of the analogical stance is still alive in the secular intellectual world.

Christian hope, as the Nicene Creed implies, denies that a flight into timelessness accurately defines human destiny. Our hope for resurrection means that our bodies, the very materiality of our existence, can be saved. And since our bodies link us to the physical universe, human hope is inseparable from the final outcome of the whole cosmic drama.

The New Testament, if read rightly, is not pointing us toward an exit from Earth and the cosmos. Christian faith longs not for the abandonment of creation but for its healing and fulfillment. Early Christian prayers, hymns, and canonical writings, especially the letters of the apostle Paul, express joy in the news that the whole of creation has become new in the person and destiny of Jesus.[3] The Gospel of John claims that in Jesus the "Word of God"—the fountain of all meaning—has now "become flesh," filling matter and time with indestructible importance. We read in Colossians that in Jesus "the whole fullness of deity dwells bodily."[4] Matter, by being the extended body of Christ, is also charged with a divine presence and promise everywhere and everlastingly.

How, then, can Christians be satisfied to look away from the cosmos or leave it behind when they die? And, as I shall be asking in Chapter 13, how can they be indifferent to ecological concern for nature here and now?

The incarnation of God in Jesus makes matter and time meaningful forever. In the Letter to the Colossians, traditionally attributed to Paul, Christians are told that the Jesus who was put to death, who was buried and is now alive, is the universe's dramatic principle of coherence, the one in whom "all things hold together."[5] In the self-sacrificing person of Jesus, Christians believe, the whole cosmic drama is joined climactically to indestructible rightness. And in the resurrected life of this defenseless victim the meaning and destiny of time are revealed for all ages.

The idea of a Christ-transfigured universe would never have taken hold, without the early disciples' experience, soon after Jesus's death, of his empowering new presence. In their earliest gatherings after his crucifixion the friends and followers of Jesus were both consoled and puzzled by the experience of their teacher's new presence among them. While some may have wanted to be carried away immediately into paradise, most of them accepted a commission to enter back into time and history.[6] They began almost immediately to spread their newfound hope over an extensive geographical area. Starting at the end of the first century, they composed Gospels to disseminate their newfound faith across the ages to come. According to these hope-filled documents, the universe is not to be "left behind" but transformed and renewed.

Interestingly, the resurrection stories also give the impression that a quality of not-yet-ness permeates the whole Easter experience. In the canonical resurrection stories Jesus encounters his friends powerfully but elusively.[7] He is with them, but he also goes "before them." He meets with his disciples but then slips out of their grasp. The disciples testify to Jesus's aliveness by proclaiming that their crucified master now "sits at the right hand of the Father" and lives on through his Spirit, keeping the future open.[8] By virtue of his "forgiveness of sins" the past no longer defines or holds power over the disciples. Witnessing resurrection empowers them to remain inside of time while also looking forward to time's fulfillment. The Nicene Creed reaffirms the hope expressed in the Gospels that a shamefully executed outcast is the one who "will come again in glory to judge the living and the dead." And it is his reign, not that of pedestrian potentates, that "will have no end."[9]

Promise

The inconclusive endings of the Gospels and the overall raggedness of the resurrection narratives may be read in three distinct ways. Archaeonomists typically debunk the New Testament resurrection stories, citing the absence of scientific evidence for the extravagant claims they make. Defenders of analogy tend to spiritualize the resurrection, forgetting the Council of Nicaea's emphasis on the inseparability of God, matter, and time, affirmed now in the Creed. The anticipatory stance, however, connects all of human history and the whole cosmic story to the resurrection-hope of the Gospels. This is because the Jesus of the Gospels—the Jesus of Christian worship—belongs to the future more than to the past or the present.[10] The earliest Christians looked for their slain and resurrected Lord to come back into the present from out of the future to renew the whole of creation. They expressed this expectation especially through the celebration of the Lord's Supper. What they hoped for was not a timeless gathering of spirits but the climax of a cosmic drama. Neither archaeonomy nor analogy is equipped to grasp or point us toward such a meaning. To understand why not, we need to probe deeper into the question of whether hope is in any sense compatible with the universe as it is understood by science after Einstein.[11]

Archaeonomy. In 1923 the British philosopher Bertrand Russell colorfully described where the universe will end up if we look at it from an archaeonomic point of view. "All the labors of the ages, all the devotion, all the inspiration, all the noonday brightness of human genius, are destined to extinction in the vast death of the solar system; and the whole temple of Man's achievement must inevitably be buried beneath the debris of a universe in ruins."[12] Such a universe, Russell went on to say, is not worthy to hold its human inhabitants.

Portraits of the universe these days do not have the same scowl as Russell's, but in almost all scientific versions of contemporary cosmology a skull is grinning in. A few scientists disagree with this pessimistic impression, including the renowned physicist Freeman Dyson, who tries hard to wipe away both the scowl and the skull.[13] Most scientists, however, claim that in spite of its having gradually given rise to thought over the course of billions of years, the universe is headed toward an abyss of mindlessness. According to archaeonomic materialists, life, thought, faith, moral aspiration—all of these fascinating outcomes of cosmic history—are reducible in the end to aimless elemental bits of matter. The great American philosopher William James, reflecting on what I have been calling archaeonomy, captures what this would mean if true:

> That is the sting of it, that in the vast driftings of the cosmic weather, though many a jeweled shore appears, and many an enchanted cloud-bank floats away, long lingering ere it be dissolved—even as our world now lingers for our joy—yet when these transient products are gone, nothing, absolutely *nothing* remains, to represent those particular qualities, those elements of preciousness which they may have enshrined. Dead and gone are they, gone utterly from the very sphere and room of being. Without an echo; without a memory; without an influence on aught that may come after, to make it care for similar ideals. This utter final wreck and tragedy is of the essence of scientific materialism as at present understood.[14]

Archaeonomy, we have seen, is unwilling to wait. It decides here and now that the universe was ontologically complete at the beginning, at the point when it was lifeless and mindless. Unlike the anticipatory stance, archaeonomists have nothing to wait for, since the cosmos was as real in the beginning as it was ever going to get. There is no room for the natural world to become more than it was at the start. Archaeonomy instructs us that what happens in the passage of deep time is that the original elements get reshuffled. What is packed into the beginning determines everything that happens later, including the eventual death of the universe.

Most contemporary scientists and philosophers subscribe to some version of archaeonomic pessimism, believing that the passage of time will have accomplished nothing in the end. In this way, archaeonomy turns out at heart to be just one more example of the distaste for time among modern and contemporary scientific thinkers. Archaeonomy insists on reducing the universe to its subatomic elements, initial conditions, and physical constants so that everything subsequent to the initial moment is anticlimactic. No room remains for fuller being, and certainly no room for hope. Archaeonomy allows for no transformative drama in which an entire universe is liberated from the deadness of the past and transformed into a story of life, thought, freedom, and faith. For archaeonomists, there is no reason to wait, let alone hope.

Analogy. Analogy, like archaeonomy, conveys little in the way of cosmic hope. It has no expectation that time is significant enough to lead to a transformed universe. Hence analogy has little incentive to look forward to what is not-yet in the flow of time. For it, the fullness of being, goodness, and beauty already resides outside the universe, in the timeless abode of an unchanging Absolute. There can be little interest in the cosmic future if everything truly beautiful and good has already been actualized in eternity. Analogy appeals, then, to otherworldly optimism but not to cosmic hope. Assuring us that each soul's destiny lies outside of time, analogy implies that after our brief stay here on Earth is over, the central core of each human being, the immortal soul, may leave the material world behind for good. Much the way Einstein's geometry melts time into space, analogical theology dissolves the vast temporal journey of the universe into the timeless dwelling place of pure spirit.

Analogy pictures our souls soaring away from time when we die, on their way to eternity, where the fullness of being is thought to reside. Analogy no doubt affirms human dignity and ennobles our personal histories, but that is because it believes our souls and personalities already belong more to eternity than to time. Quite early in Christian history, variations on the analogical worldview turned the religious longings of the faithful toward a spiritual world beyond matter and time. Analogical Christianity ever since has looked for the fulfillment of Jesus's promises not in a transformed universe but in a harvesting of souls from the universe.

Analogical theology fails, therefore, to make resurrection-hope relevant to the question of cosmic destiny. For analogical Christians the drama of redemption is sometimes mixed with elements of biblical anticipation, but in its usual forms, analogical theology reduces Christian hope to the salvation of souls. During the early Christian centuries, the claim that Jesus was alive began to drift away from its native Abrahamic anticipatory moorings, eventually coming to rest in a neo-Platonic piety that longs for the core of our human existence to be transported out of time and into the spiritual splendor of eternity. Interpreted analogically, the Nicene Creed's profession of belief in "the resurrection of the dead" and "the life of the world to come" has usually meant the interruption of time by eternity rather than the fulfillment of time in the compassion of an incarnate God.

Anticipation. New and unpredictable things have been happening in our universe for billions of years, including the recent arrival of inquiring minds. So there is no reason to assume that the cosmos has yet run its course in irreversible time. Quite probably, more surprises will be arriving from out of the future. If so, to understand the universe, we have to keep scanning the horizon up ahead in search of a still-out-of-sight dramatic coherence that can be encountered only by those willing to wait in hope. Instead of digging back into the universe's granulated opening moments or springing headlong into eternity, anticipation stays close to the cosmic ground. It looks not up above but up ahead for a dramatic coherence. From the perspective of anticipation, Christian resurrection-hope cannot be separated from hope for the future redemption of the entire universe.

What, then, are we to say about the elusiveness, or the not-yet quality, of the biblical resurrection accounts? I think we may read the Gospel stories with the same anticipatory expectation with which we have been reading the unfinished universe all along. In the New Testament the resurrection is not a climactic epiphany of God but a promissory event that points Jesus's disciples in the direction of a new future. The narrative form of the biblical accounts of Jesus's resurrection is different from reports of visitations by the gods in pagan religions. It is more akin to the ancient biblical narratives of the promissory appearances of God in the stories about Abraham and Moses. In these ancient accounts God appears, makes promises, elicits the response of faith, and then disappears into the future. The New Testament stories about the resurrection of Jesus have a similar quality. If we read these stories in an Abrahamic anticipatory spirit, we notice that the Easter event resists being pinned down to the present or shoved into the past. Jesus goes before or ahead of his earliest followers, calling them toward a new future.[15] We may wonder whether his disciples would have had any resurrection experiences at all had they not already been steeped in the habit of anticipatory hope by their familiarity with Jewish spiritual traditions going back to the ancient stories about Abraham.

Anticipation, I want to emphasize once again, is completely compatible with the methods and discoveries of the natural sciences. It endorses analytical and mathematical inquiry because these are essential to grasping the physical constraints that grammatically hold together the moments and episodes of the drama of awakening that we call the universe. I have not been criticizing science, whose practice requires great patience, but archaeonomy, which has a built-in epistemological impatience. Likewise, I have criticized the analogical vision not for its compassionate longing for an end to suffering and death but for its refusal to incorporate the whole cosmic story and the evolution of life into its optimism. I have taken issue with both the analogical and the archaeonomic stances because, instead of waiting for the universe to reveal its meaning up ahead, they declare dogmatically that the fullness of the world's being lies either in eternity or at the beginning of cosmic history. Neither hope nor moral aspiration, I believe, can survive for long in either milieu.

Fulfillment

Christian hope implies that time has a meaning and that it carries a divine promise of final fulfillment. But the universe as science sees it is destined eventually to collapse physically and energetically. How can we hold these two readings—one of promise, the other of perishing—together? Is the whole long story of the universe destined to be lost and forgotten completely in the end, or can it somehow be remembered forever?

Let me begin a response by noting that human memory, using the tools of science and history, can recall past episodes of the cosmic story long forgotten. It is the function of memory to give past events new life in the present, but it is still a mystery that we can talk about the past at all. If time makes every moment perish, then being able to refer to what happened earlier in time implies that the past has not perished absolutely. All the series of moments that have made up the cosmic story are still around, somehow waiting to be recalled. Where are they waiting?

Christian theology submits that the ultimate repository of past events, including the whole of cosmic history, is the infinite care and compassion of God. In fact, most religions are attractive to their followers because of their promise, expressed in thoughts about immortality, resurrection, and reincarnation, that the past can never be fully forgotten. The philosopher Alfred North Whitehead has developed an interpretation of the cosmos in which events that took place in the remote cosmic past are still resonating in each present moment. Many fields of research, ranging from the neurosciences to astrophysics, may also contribute their own answers to the question of how remembering takes place. Theology itself fully supports scientific research into the mysterious power of memory. But since the universe can be read at different levels, as I have been saying from the start, there is room for both scientific and theological ways of making sense of things, including time and memory.

That God never forgets is a fundamental belief of Abrahamic religion. This teaching can be consoling, however, only if God's remembering is not a mere accumulation of information but above all an exercise of divine compassion. Recall, for example, Jesus's message that the very hairs of our heads are numbered, or the psalmist's cry: "You have kept count of my tossings; put my tears in your bottle. Are they not in your record?"[16] Christian hope means that every moment of experience, both human and nonhuman, is rescued and saved everlastingly in the breadth and depth of divine compassion—and with full experiential immediacy.

The Protestant theologian Paul Tillich captures the classical Christian sense of divine memory in terms that do not differ altogether from those of Spinoza and Einstein:

> Nothing truly real is forgotten eternally, because everything real comes from eternity and goes to eternity. . . . Nothing in the universe is unknown, nothing real is ultimately forgotten. The

atom that moves in an immeasurable path today and the atom that moved in an immeasurable path billions of years ago are rooted in the eternal ground. There is no absolute, no completely forgotten past, because the past, like the future, is rooted in the divine life. Nothing is completely pushed into the past. Nothing real is absolutely lost and forgotten. We are together with everything real in the divine life.[17]

If time means an irreversible passage from past to future, however, Tillich's analogical sense of divine care needs to be expressed in a more anticipatory way. The fundamental units that make up our time-ferried universe are not atomic physical bits, as Democritus and modern atomists may have thought. Rather, nature is made up of temporal moments. If time is real, then the basic constituents of our dramatic universe are not spatial objects but unrepeatable events.[18] Each event in cosmic process is a singular throb of existence that is fleetingly actual and then perishes. But it does not perish absolutely. The temporal events that occur in the cosmic story are received and transformed in subsequent events. Traces of the entire cosmic past are retained in every present moment, so there is no absolute loss in any temporal series. Occurrences or events, even though they perish individually, keep adding up or accumulating in the irreversible flow of time. Events do not dissolve into nothingness as the cosmic drama unfolds. In some way, past moments are still extant, or else we could not refer to them or talk about them at all.

But, again, where are the past moments that make up the cosmic story waiting? Whitehead, who was familiar with the work and person of Einstein, took this question seriously. He agreed with Einstein that time is inseparable from nature, but contrary to Einstein, he insisted that time is real. Time is an irreversible passage that cannot be fully grasped by geometry. Indeed, Whitehead thought that the modern obsession with geometry was partly responsible for the loss of a sense of real time on the part of scientists and other thinkers. Whitehead was critical of Einstein's cosmology, therefore, for its failure to differentiate past, present, and future. He must have noticed that Einstein had associated God with the timeless geometry of the universe, thus blunting the cutting edge of real time and nullifying the experience of real perishing.[19]

Einstein dealt with the fact of perishing by endowing the universe's geometry with the attribute of timelessness, wherein there can be no real loss. Consequently, the analogical side of Einstein's philosophy of nature led him, Platonically, to identify the real universe with the eternal geometric perfection that "tells matter how to move."[20] Since what happens in time is always subject to loss, the place for humans to hang out, it would seem, is not in time but in eternity. In his passionate love of timeless geometric perfection Einstein was no less religious than most other devotees of analogy. He assumed that time cannot be real in comparison with the eternal mathematical perfection with which nature is organized.

Nor does Einstein's idea of God, unlike Whitehead's, require the attribute of personal caring. For Einstein (like Spinoza), it was almost as though God, whom Spinoza identifies with nature, is too magisterial to condescend to caring for what happens in the lowly realm of time-passing. For Whitehead, time and perishing were real, and God was deeply personal and temporal. God, he said, is "a tender care that nothing be lost."[21]

Not only is time real but so also is perishing. The main problem for philosophy, religion, and theology, then, is to find a way to respond to the fact that things in time inevitably perish. In Whitehead's way of thinking, Einstein did not take the passage of time seriously, so he failed to take the fact of perishing seriously as well. Where there is no real loss, there is nothing to save. God, according to Whitehead, is inseparable from time and loss, and all events are rescued from absolute perishing by the same responsive God who offers relevant new possibilities to the creative cosmic process. This God is the ultimate ground of our hope, and there is no genuine hope apart from the redemptive compassion of God.[22]

In Whitehead's theological cosmology even though each moment in time perishes, it is felt fully by God and remains alive in God's experience—cumulatively and everlastingly. Everything that has ever occurred in the cosmic story is received into God's life without fading—that is, without the loss that we ourselves experience when things we cherish are gone. God's preservative care is the ultimate reason why the past has not perished absolutely and why we are still able to recall it and ask about its meaning. Historians, in their concern to know what has happened in the past, are unknowingly instruments of the everlasting divine care that nothing be lost.

In Whitehead's view, God is the ultimate, infinitely caring recipient of all the moments of experience that make up the temporal world.[23] God retains all events in increasingly intense feeling, thus redeeming everything that happens in the cosmic drama, including the suffering of living beings. But, as Whitehead argues, there is more to God than just preserving and saving the past. I take this to mean that God is also the ultimate reason why nature is a story. The temporal universe, as I have been saying, is not just the predictable outcome of timeless geometry and eternal laws of nature. It is also a dramatic awakening that wanders far outside the boundaries of geometry. To have the shape of drama—or what Whitehead calls "adventure"—the universe, at each moment of its existence, must be open to new possibilities that arrive unpredictably in a way that no amount of scientific expertise could ever fully anticipate.

God is the ultimate reason why new possibilities exist. God—as I interpret Whitehead—is the ultimate source of the novelty that makes the universe dynamic and dramatic.[24] God transcends the universe not only spatially but also temporally. God, I have been proposing, is the not-yet that keeps the future open and allows room for new possibilities to greet the past passage of time and give it continually new meaning in each present moment. God is unsurpassably intimate with every moment that makes up the cosmic story. As the cosmic passage of time is taken into the divine life, each moment is related in a novel way to the ever-expanding pattern of beauty that already makes up God's inner life. This, too, is a reason for hope.

The point of the universe, according to Whitehead, is the building up of beauty.[25] Pope Francis seems to agree when he writes in his encyclical on ecological responsibility that "in the end we will find ourselves face-to-face with the infinite beauty of God (*cf.* 1 *Cor* 13:12), and be able to read with admiration and happiness the mystery of the universe, which *with us* will share in unending plenitude."[26]

Even though from a scientific perspective the universe will eventually undergo death by energy collapse, the theological position summarized here allows that everything that goes into the cosmic drama is everlastingly sublimated, preserved, and redeemed in God's own compassion—that is, in God's capacity to care forever. All that goes on in time, therefore, is not lost but preserved and transformed by God's increasingly widening vision and memory into a depth of beauty that promises to redeem all suffering and loss. In this theological understanding, God suffers and strives along with the world. Hence, it is not the timeless perfection of geometry but the limitless compassion of God that gives us reason to hope in the face of time's perpetual perishing.

This account notwithstanding, the question can be asked again: Is Christian resurrection-hope anything more than a comforting illusion? Our contemporary archaeonomic intellectual culture will inevitably reject every theological vision, including the one that I have all too succinctly summarized just now. Let us recall, however, the distinction made here between the universe's geometric and dramatic coherence. Einstein's relativity physics provides a good example of the geometric coherence that ties the universe together gravitationally. Christian faith, however, claims in effect that in the resurrection of Jesus the whole universe reaches—by anticipation—what I have been calling a dramatic coherence.

In order for the universe to be the carrier of dramatic meaning no violation of nature's physical regulations occurs. The universe can undergo dramatic transformations that do not show up on maps of geometric understanding. We have already seen an instance of this nonintrusiveness in the surprising appearance of the first living cells on Earth 3.7 billion years ago. From the point of view of physics and chemistry, no habitual routines in the physical universe had to be interrupted to let life come into the story. But even though the origin of life required no suspension of physical and chemical routines, the universe by virtue of that event suddenly became completely new, dramatically speaking. What could be more dramatic, after all, than the story of a lifeless universe becoming a living one?

The drama of life entered into the universe without making a crease in its physical and chemical patterns or in its geometric coherence. Consequently, for Christians the dramatic events at the root of their sense of a renewed universe are trivialized if we try to confirm them by way of the same method of inquiry we use for finding geometric coherence. It is by way of anticipatory hope, not by science or mathematical abstractions, that dramatic coherence is encountered by those who are prepared to wait. Neither archaeonomy nor analogy has the requisite epistemological patience to encounter the universe in the fullness of its being and becoming.

Summary

Theology, I have been saying, looks to find reasons for our hope. As a general rule, theology in the age of science does not look for reasons to hope by pointing to miraculous events in the habits of nature, as Einstein assumed. In anticipating life's final victory over death and expecting "the life of the world to come," theology may read the same cosmic story in different ways. One way is that of analogy; another the way of anticipation. These two readings sometimes merge and mix in the minds of individual Christians, but analogy has usually overpowered anticipation. And both readings have had to combat a third, the fatalism and cosmic pessimism that today finds its home in archaeonomic depictions of nature.

Christians, in past ages, have looked at both the Nicene Creed and the natural world mostly with an analogical eye. During the fourth century, when the articles of the Nicene Creed were being painfully put together, its architects favored the analogical stance. They were unable at the time to connect their resurrection-hope to irreversible cosmic time and an awakening universe. Part of the reason for the analogical emphasis in so much Christian spirituality lies in the wording of the Creed itself. Reflecting biblical imagery, the Creed professes that after Jesus died and rose again, "he ascended into heaven." The ascension into heaven has often been pictured analogically to mean that a savior has finally opened up an avenue from time to eternity. The expectation that Jesus will "come again . . . to judge the living and the dead" suggests that the saved are destined to be snatched from the temporal world and transported into the spiritual world of timeless perfection up above.

This, however, is a misrepresentation of the dominant meaning of Christian hope, as the Second Vatican Council (1962–1965) recently recalled, centuries after Nicaea.[27] The Jewish background of early Christian theology looked forward to the renewal, not the abandonment, of the cosmos. The Christian resurrection-hope originally entailed the revitalizing and transforming of creation, not an escape from the cosmos. The biblical doctrine of creation and the New Testament proclamation that the Word of God "became flesh" are assurances that matter and time have always been theologically important. Time, in other words, matters to God. The incarnation of God, reaffirmed emphatically at Nicaea and again at Vatican II, means that matter and time are everlastingly inseparable from the life of God. This implies, I believe, that what happens in time contributes something to the very identity of God. Just as a loving parent cannot help being changed by the suffering and joys of her or his child, so also a caring God cannot help being affected—and that means transformed in some way—by what happens in the cosmic story.[28]

Finally, although the universe is tied at one level to unbreakable physical and chemical routines, it reveals itself at another level as an unpredictable drama going on in deep time. It is especially in the drama rather than the geometric grammar of the universe that we look for reasons to hope. And even if the universe is condemned physically to eventual death, whether by heat or by cold, the trail of moments that are making up its dramatic interior do not end up vanishing into a void. The story of the whole universe, Whiteheadian theology speculates, is registered permanently within the hidden, indestructible rightness to which the universe is awakening. Christians hope also that the imprinting of the whole cosmic story on the compassionate "memory" of God also includes in some way the subjective survival and ongoing transformation of personal consciousness after death.[29]

TWELVE

Compassion

> A man's ethical behavior should be based effectually on sympathy, education, and social ties; no religious basis is necessary. Man would indeed be in a poor way if he had to be restrained by fear of punishment and hope of reward after death.
> —Albert Einstein

> Earthen natural history might almost be called the evolution of suffering. But that makes it equally plausible to call it the evolution of caring.
> —Holmes Rolston III

FOR MORE THAN TEN BILLION years our universe slept silently. Then, around 3.7 billion years ago it came alive in the form of single cells. For a couple more billion years life remained fairly simple, but around 570 million years ago, during the Cambrian epoch, it began to become more and more complex at an accelerating pace, giving rise to multicellular organisms that eventually evolved into sentient subjects. Subjects are living centers of experience that in the course of cosmic time have gradually become conscious and curious. An increase in neurological complexity made possible the gradual emergence of minds and a dramatic leap into thought during the past few million years. In beings endowed with minds—at least on Earth and perhaps elsewhere amid the many billions of galaxies—our universe gradually gave rise to moral aspiration and, in the case of religions such as Buddhism and Christianity, to an unprecedented cultivation of the virtue of compassion. The dawning of compassion, I suggest, like the rise of life and thought, marks a distinctively new phase-change in the universe's long dramatic awakening.

Formerly nobody would have thought of the birth of compassion—or any other manifestations of what we call virtue—as a cosmic development.

Until recently even the greatest religious thinkers and philosophers of nature had no awareness that the universe itself is the subject of an awakening and that matter could undergo a transformation into such surprising outcomes as life, mind, freedom, faith, hope, and love over the course of time. The physical universe was usually thought of solely as a staging ground for the story of life and eventually of human history. It did not occur to the ethical thinkers of the past, such as Immanuel Kant, that moral aspiration is a new development in cosmic history and not just a refinement of human subjectivity.

In this chapter, I want to look into the cosmic meaning of compassion. I see the arrival of compassion in the cosmic story—and proximately in life's evolution—as emblematic of the whole list of high virtues idealized by people of faith and cherished by moral philosophers over the past 2,500 years. After Einstein it is possible, for the first time in the history of thought and spiritual life, to interpret compassion not only as a sensational step in the moral development of human beings on Earth but also as an important new stage in an awakening universe.

We now know that the universe has never remained continually the same. It has been marked throughout by phase-changes in which unpredictable outcomes have burst onto the cosmic scene, though often quietly at first. In the human species' cultivation of the virtue of compassion we witness the arrival of one of the most surprising outcomes of the cosmic story so far.

During the past two centuries geologists have mined from the earth an informative record of fossils. The distinct layering of the fossils, from lower to higher levels, tells the long story of matter becoming more and more complex. Now the history of nature tells us that complexity became, over time, increasingly vitalized, eventually sentient, conscious, self-aware, and, fairly recently, attracted to goodness, truth, and beauty. During and after the invention of symbolic thought in human evolution the cosmos became not only conscious but also, at least occasionally, compassionate. Evidence of the early emergence of compassion among humans shows up at gravesites—for example, in prehistoric burial practices that included the ornamenting of bodies. Such rituals point to an early human concern for the destiny of the dead that could never have taken root apart from an intensification of compassion in the hearts of the living. Not only sadness at the death of loved ones—as occurs also in some other mammalian species—but also empathy for those who have died has been a motivating factor in our ancestors' religious imagining of new life beyond the portal of death.

Along with the evolution of compassion among humans there arose an awareness, expressed in the narrative form of what we now call myths, that the world is not right in some sense as it presently exists. Human awareness of the gap between the actual and the ideal has come to expression in countless religious myths about the origin and end of evil. These myths have envisaged a realm of rightness that exists indestructibly beyond the wrongness of pain and death. The Nicene Creed displays its continuity with this ancient religious sensitivity when it professes hope for "the resurrection of the dead and the life of the world to come." Most religions insist that wrongness will finally be defeated by rightness. Redemption is possible, but we may have to wait for it—and not without being tempted at times to despair.

In human experience everything right is mixed up with some degree of wrongness. How is this possible? Wrongness, one may reply, is likely to befall any universe that has yet to become fully real. As long as the cosmos is still in the process of being born, and perhaps of becoming more, each present moment reflects in some way the overall incompleteness and imperfection of the cosmic story. The present inconclusiveness of a universe still coming into being inevitably leaves open a space for wrongness, both natural and moral.[1] Natural wrongness—the evil built into nature long before humans came along—includes not only the disintegration and perishing of organisms but also the excess of suffering and predation that accompanies the story of life. Moral wrongness, the kind of evil that humans have introduced into the world, consists of destructive acts and omissions that run contrary to the ongoing awakening of the universe. In the context of a still-emerging universe, moral evil—or what the Abrahamic traditions call sin—is our free human refusal to participate in the dramatic enhancement of life, mind, faith, freedom and hope. Sin, cosmically speaking, is a deliberate refusal to let these emergent phenomena survive and prosper into the future.

The Evolution of Compassion

To morally sensitive and scientifically educated humans, the world has never been completely right. Evolutionary scientists, for example, are especially disquieted at the suffering of life during the many millions of years after organisms acquired sensitive nervous systems. Just as Charles Darwin protested the excessive suffering of life, so also do contemporary compassionate biologists. Many who formally espouse archaeonomic materialism have denounced the inherent wrongness of evolution for its indifference to suffering. These sensitive scientists are morally disturbed at how the universe brings life into being by way of the crude mechanism of natural selection.[2]

Darwin, during his studies at Cambridge and prior to his famous journey on the HMS *Beagle* (1831–1836), embraced the traditional religious belief that the natural world somehow reflects the transcendent goodness of God, albeit imperfectly. As a young man, he accepted the arguments of Christian natural theology that nature is the product of beneficent divine design. Later on, however, he came to realize that life is a long and often violently creative process, one in which individual organisms suffer excessively and then perish. Darwin's unusually strong attraction to the virtue of compassion made him wince at the way life works. A constant source of anguish to his delicate conscience was that the cruel mechanism of natural selection is an indispensable causal agency in the creation of new species and the story of life. Life, as humans have always known, is often harsh, but prior to Darwin the full scope of life's suffering had never been so rawly revealed. The organic design that had earlier seemed to be the product of kindly divine governance turned out, in Darwinian perspective, to be the outcome of a long series of accidents combined with the heartless process of selection. Sadly, the evolutionary process leaves organisms and species adrift in a stream of undeserved suffering. And in the case of most species of life, evolution has led to the fixed state of extinction where they now rest, apparently unredeemed and unremembered.

Our present awareness of the engineering inefficiency of evolution, combined with the pain, death, and wasteful amount of time it takes to bring about life's relatively few survivors, makes it hard for many ethically idealistic people to attribute rightness to the universe. Thus, the thoughtful philosopher Philip Kitcher has exclaimed that "a history of life dominated by natural selection is extremely hard to understand in providentialist terms." Nature is inseparable from wrongness, as Kitcher and many other Darwinians see it. "Indeed," Kitcher comments, "if we imagine a human observer presiding over a miniaturized version of the [evolutionary process], peering down on his 'creation,' it is extremely hard to equip the face with a kindly expression."[3]

We have already situated the evolution of life, mind, freedom, and faith, however, inside the long cosmic narrative laid out by science since Einstein. May we not also do the same with the recent emergence of compassion? A cosmic perspective allows us to understand in a new light almost everything we had previously located inside a fixed universe. An anticipatory stance, instead of focusing only on the wrongness of suffering in the story of life, is aware that the awakening of human beings to compassion is also an outcome of our long cosmic story. The sentiment of compassion that leads Darwin and other fair-minded evolutionists to protest the excessive suffering of life is just as much a part of an awakening universe as the heartless mechanism of natural selection has been.

So, while cringing with Darwin at the insensitivity of the evolutionary process, let us, first, take into account the new post-Einsteinian scientific awareness that all living beings and indeed the whole of life's evolution are part of a much larger cosmic drama. Second, let us recall that the drama is one of gradual awakening. Third, let us observe that this awakening includes, at least in one newly conscious species of life, an increasingly refined sense of the distinction between wrongness and rightness. And, fourth, let us not ignore the fact that the same universe that has sponsored the wasteful process of natural selection has also given rise lately to beings who are exceptionally sensitive to the rightness of compassion and the wrongness of life's excessive suffering.

If the cosmos itself is undergoing a long and gradual awakening, we are permitted, as we read the story up to the present, to keep our eyes not only on the insensitivity of evolution in the past but also on the rise of compassion more recently in the evolution of human consciousness. They are both part of a single cosmic drama. Darwinian science has brought to light what compassionate humans take to be nature's past wrongness. But the cosmic drama, we need not forget, is much more than its past and present products. It is also in some sense not-yet. What the universe is really all about has yet to be fully revealed.

Moreover, it is good to remember that natural selection has never been unambiguously evil. Suffering itself, as Darwin knew, is an adaptive trait that contributes to evolutionary fitness. It is true that evolution has been a narrative of competition, struggle, and strife, but it has also been a tale of emerging creativity, cooperation, and care. Even if the story offends the sensibilities of beings who aspire to moral rightness, it must not be forgotten that evolution has also created the genes and nervous systems of those very human organisms that are now protesting what they take to be the wickedness in nature. The recent arrival of compassionate beings in the universe is not an insignificant afterthought in the long—and still unfinished—cosmic story. Our human awakening to the ideal of compassion is no less integral to the cosmic story than are the impersonal natural processes that gave rise to living and thinking organisms long ago.

To archaeonomic naturalists, however, any suggestion that their own sense of moral rectitude is part of a long process of cosmic awakening will, no doubt, sound strange. Once again, though, this impression occurs because most archaeonomists, as I have observed several times already, do not really believe that their own minds and moral aspirations are part of the universe. Archaeonomists still adhere unknowingly to the Platonic assumption that their own intelligent and moral subjectivity exists somehow apart from the objective world. Some of them doubt that moral subjectivity, at times even their own, has any real existence at all.

The paleontologist Stephen Jay Gould, in his laudatory compassion for the suffering of living beings, exemplifies the typical archaeonomic assumption that human moral consciousness is not really part of the universe:

> When we thought that factual nature matched our hopes and comforts . . . then we easily fell into the trap of equating actuality with righteousness. But after Darwin . . . we finally become free to detach our search for ethical truth and spiritual meaning from our scientific quest to understand the facts and mechanisms of nature. Darwin . . . liberated us from asking too much of nature, thus leaving us free to comprehend whatever fearful fascination may reside "out there," in full confidence that our quest for decency and meaning cannot be threatened thereby, and can emerge only from our own moral consciousness.[4]

Gould, like many other archaeonomic thinkers, fails to notice that his own moral consciousness belongs fully to an awakening universe and is in fact an especially important instance of that awakening. In the passage just cited he tacitly situates his moral sensitivity "in here" apart from the universe "out there." He ignores the fact that his own compassionate moral life belongs to the very same cosmic story as the mechanism of natural selection and the struggles of prehuman life. Unfortunately, in the very act of condemning natural selection and the whole universe for their cruelty, Gould forgets that his own sense of moral righteousness is an outcome of the same long cosmic awakening that earlier gave rise to life. If, as I have been arguing, human mental and moral life are integral to the cosmic awakening, it follows that we can learn at least as much about the universe by contemplating the newly emergent human capacity for compassion as we can by looking at instances of evolutionary cruelty.

By acknowledging that rightness has dawned in the universe only gradually, theology after Einstein does not have to suppress awareness of the morally troubling Darwinian chapters. These are part of the same big story that has led, not too long ago, to our species' refined awakening to the virtue of compassion. Scientifically speaking, the cosmos has often undergone dramatic transformations in the course of deep time. It is still awakening—hesitantly and not without setbacks—to rightness. From a narrative point of view, the newly emergent sensitivity to life's suffering on the part of Buddhists, Christians, and morally aware Darwinians, among many others, belongs to the same cosmic drama as the seemingly cruel grammar of natural selection.

The remarkable phenomenon of human compassion, in other words, must not be left out whenever we ask what the cosmic story is all about. It is characteristic of archaeonomic mythology, however, to read natural processes by looking at their earliest stages and ignoring later developments as incidental. An Abrahamic cosmic sensibility, by contrast, fully acknowledges the past and present imperfections of nature, including design flaws and evolutionary suffering. The anticipatory stance makes room for cosmic meaning and redemption by looking toward a future coherence presently out of sight. Since the universe is still aborning, a cosmically informed faith protests the evil of suffering in evolution, of course, but it does not for that reason impatiently condemn the whole cosmic process, especially since its dramatic awakening is not yet at an end. The bitter protest against life's suffering by morally sensitive evolutionists is itself a phenomenon that we may now look at cosmologically. Their compassion is telling us something not only about their personal character but also about the universe.

Impatience

The archaeonomically minded Darwinian pessimists, we note once again, are unwilling to wait. Even though the cosmic story is not over, they decree with unquestioning certitude that because of the immorality of natural selection, the universe is pointless, evil, and absurd. They tell us that any world created by God should have been put together perfectly from the start. Yet a world fully fashioned at its inception would not be one into which compassion could ever conceivably have entered. It has been in the crucible of life's suffering that compassion has emerged. Evolutionary naturalists such as Kitcher are in effect insisting on an initial instantaneous perfection as a condition of their accepting a providential interpretation of the universe. Suffering should never have been part of the cosmic story, they believe. But the fully perfected initial creation they have in mind would have left no interval for a temporal transition between beginning and end. It would have allowed no opening for a drama that might carry meaning. A world magically ordered to timeless perfection from the start would have allowed no clearance for the passage of time, the emergence of complexity, the striving of life, the exercise of human freedom, and the sublimation of sentience into the splendor of compassion.

For that reason, as I have argued throughout the preceding pages, Christian faith after Einstein makes more sense if we detach it from its ancient and medieval analogical environment and return it to the anticipatory setting of Abraham, the prophets, and Jesus. To do so, we need to learn once again to cherish the full reality of irreversible time. Accepting the passage of time as real is not only theologically justified by its fidelity to Abrahamic faith, but it also brings Christian faith into a more meaningful encounter with cosmology after Einstein, Lemaître, Hubble, and Hawking. The universe of contemporary science is a long story whose intelligibility cannot reveal itself in the stiffness of mechanical design or the reverie of analogical metaphysics, but only in the indeterminate fluidity of a narrative whose full coherence has yet to arrive.

Steeped for centuries in time-despising visions of the natural world, analogically influenced Christian theology was not ready for the unfinished universe disclosed by contemporary biology and cosmology. It still struggles to realize that nature cannot be an immediate instantiation of rightness without forfeiting its narrative capacity to carry meaning and open itself to being more. Neither the wrongness of evolutionary suffering nor the unspeakable evils in human history can ever be rendered intelligible in terms of any present understanding of things. I am aware of no scientific, religious, theological, or metaphysical system that at this moment can make good sense of suffering and death. And even if a system of thought claimed to do so, it would subtly legitimate wrongness by giving it a seemingly intelligible place in the total scheme of things.

Wrongness is not a topic for reason but a wound calling for healing. Because of Christianity's expansive hope for the ultimate termination of moral evil, suffering, and death, Christian faith, as I understand it in light of the Nicene Creed, stands in real, not just theoretical, contradiction to the suffering of life. This is because its resurrection-hope looks not for an escape into timelessness but for a fulfillment of time in the compassion of a time-loving God. In this reading, human acts of compassion and healing that we witness here and now are installments—or anticipations—of a final dramatic coherence in which every tear is wiped away and death is no more.

The Problem of Evil and the Reality of Time

How can Christians believe in God if natural suffering and moral evil are also allowed to exist? If God is perfectly loving and infinitely good, why does God tolerate suffering and other kinds of evil? The "theodicy problem"—how to reconcile the existence of God with the fact of evil—has tormented generations of philosophers and theologians, but it has done so almost always in the context of an analogical reading of nature. Seldom, if ever, have conversations about the problem of evil started out with the anticipatory premise that God is somehow not-yet. Furthermore, those who have wrestled with the problem of evil and suffering have wrongly associated perfection with timelessness, and evil with time. The anticipatory theological stance is not obliged to follow these cramped definitions.

In our anticipatory understanding, the extent of God's goodness is not measured by distance from what happens in time. Divine perfection does not consist of abiding in timelessness, apart from all pain and perishing. Rather, it consists of a capacity for unsurpassable intimacy with the passage of time and the suffering of life. God is not an "eternal now," detached from time and matter, but a compassion that gathers up, heals, transforms, and saves everlastingly whatever happens in time.

If, with analogy, we assumed that perfection means timelessness, the changeless divine presence would leave no room for a creation distinct from God, or for a cosmos with an open future, or for human persons endowed with freedom. Divine power, in our anticipatory understanding, is the unrestricted potentiality of the not-yet. A dramatic universe could not have been polished to the point of being finished in an initial act of divine magic. Moreover, an initially finished, perfected, and timeless universe would be devoid of freedom, since everything would be fixed in its place forever. An instantaneously finished creation would have been devoid of a future, since there would be no series of moments making the transition from before to after. An initially fixed and finished universe would also be devoid of life, since there would be no opening for the anticipatory striving that is essential to the definition of living beings. Finally, an initially fixed and finished universe would also be devoid of meaning, since it would not allow for the irreversible passage of time essential to the gradual emergence of narrative coherence.

Evolution and contemporary cosmology, by making life and matter inseparable from time, have destroyed in principle the religiously literalist belief that the world has been complete from the beginning. By contrast, both analogy and archaeonomy prefer a universe that was finished at the start. Both analogy and archaeonomy harbor an implicit wish that the universe not be permitted to be a dramatic story of awakening in irreversible time.

Defenders of archaeonomy boast that they are just being realistic, but they fail to see that the cosmos has, at least so far, always left open a wide swath of narrative space for the arrival of fuller being, more intense life, deeper forms of beauty, and higher degrees of moral goodness. The lifeless universe of 13.8 billion years ago would have looked unpromising to anyone who could have observed it that early. Yet, as science has now shown, the cosmos was open even then to the eventual coming of such splendid outcomes as life, thought, freedom, and compassion. Here and now, as we look back over the cosmic story as it has taken place so far, how do we know that, in spite of the darkness still surrounding us, there may not be room for fuller awakening and more illumination up ahead?

Defenders of the classical analogical way of dealing with the problem of evil likewise remain mostly out of touch with the post-Einsteinian temporal universe. Their ideal of an initially completed creation implies, after all, that the passage of time and history can contribute nothing to the increase of being. To analogy, the long suspenseful story of nature seems irrelevant to the genesis of moral aspiration. The analogical stance, no less than the archaeonomic, finds no moral or spiritual significance in the long passage of time. It allows that the practice of virtue enhances our interior lives, builds human community, and even leads to happiness, but it sees no connection between the birth of compassion and the drawn-out drama of an awakening universe.

The archaeonomic assumption that the virtue of compassion is merely a product of genetic determinism and evolutionary selection renders morality meaningless, whereas the analogical assumption that perfection has already been fully realized eternally is morally dispiriting. If, on the other hand, we acknowledge that creation is not yet over and done with, that time is real and God is not-yet, then there is plenty of room for effective human moral action, along with a sense that the human species—in its capacity for growth in compassion—has an especially important cosmic vocation.[5]

Summary

The classical prescientific analogical vision assumes that creation was perfectly good and complete in the beginning. It also assumes that the world of time and matter, to which we have been exiled, is estranged from the timeless heavenly world that is supposed to be our true home. Viewed analogically, the life of virtue in our temporal world is a matter of remembering the realm of eternal goodness from which we are now estranged.

The passage of deep time, in the analogical reading of the cosmos, is pointless and undramatic, since the universe was supposedly created complete in the beginning. Cosmic and human history are rendered inconsequential, nor can our moral lives contribute any substantive enhancement to the universe. The analogical assumption that perfection has already been fully realized eternally can only be morally deflating, since everything important will have been accomplished already. Ignoring the prospect of a truly unprecedented cosmic future, the analogical reading of the universe leads us to wonder what the full meaning of compassionate moral action could possibly be here and now.

In the light of an anticipatory cosmology an initially perfected state of cosmic being exists only in our minds and imaginations, since it has never yet been actualized in time. In an unfinished universe there is always space for both natural and moral evil. The fact of evil shows up presently not by contrasting time to eternity but by noting the dramatic temporal gap between what is actual and what is anticipated. As long as the cosmic drama of awakening is still going on, the universe remains at least partly in the dark. An awakening universe inevitably has a shadow side. Natural wrongness, as we have now learned from science, was part of the story of the universe and of life long before we humans came along. Wrongness can slip into any process that remains short of rightness. As long as the cosmos still awaits complete communion with the boundless compassion of God, there remains room for wrongness. There remains also room for hope and redemption.

THIRTEEN

Caring for Nature

> At the end, we will find ourselves face-to-face with the infinite beauty of God (*cf.* 1 *Cor* 13:12), and be able to read with admiration and happiness the mystery of the universe, which with us will share in unending plenitude.
>
> —Pope Francis

ALL OVER THE EARTH, AND especially in the most impoverished lands, sources of fresh water are drying up, forests are being cut down, topsoil is washing away, and deserts are spreading. Earth's atmosphere and oceans are being polluted, the planet's climate is undergoing a dangerous transformation, and species of life are disappearing for good at an alarming rate. Widespread warfare, excessive consumption, and burgeoning human populations have been adding to the stress that the human species is placing on the life-systems of our terrestrial home.

Undeniably, something has gone terribly wrong with humanity's relationship to the natural world. Because of the immense scale of environmental damage, it may be tempting to approach this state of affairs in a spirit of pessimism. Or, for those who are analogically motivated, it may be comforting to take refuge in the assumption that the natural world was never intended to last forever anyway. If religious faith is supposed to prepare souls for a better home elsewhere, it is even possible to ignore our ecological predicament altogether.

Or, instead, as I shall argue here, we may treat the relatively new human trend of consciously caring for nature as an essential new stage in the drama of an awakening universe.[1] In our own species the universe not only becomes conscious of itself but may also begin—more consciously than ever—to take care of itself.

The "Laudato si'" encyclical of Pope Francis is one among many encouraging signs that Christians are beginning to experience a new relationship with the natural world. Our caring for nature is not simply a matter of saving ourselves and other living beings, or of ensuring the fertility of life, or of practicing faithful stewardship in obedience to God. All of these are good reasons to care, of course, and Christian theologians are right to keep looking into the Scriptures and religious classics in search of a doctrinal foundation for supporting the ecological movement. But is that enough?

Over the past thirty years or so, ecological concern has led Christian theologians to the surprising discovery that religious incentives for ecological responsibility have been present all along in the Bible, in time-honored sacred literature, and in sacramental practice. Christian theologians are beginning to appreciate the ecological relevance of a biblical valuation of creation, with its emphasis on the intrinsic goodness of nature, the importance of stewardship, and the necessity of Sabbath rest to mitigate our temptations to exploit the gift of life selfishly. Our new situation has brought out the ecological importance of biblically sponsored virtues such as humility, temperance, justice, gratitude, compassion, and hope. Christian theology can also now discover in classical theological literature a wealth of previously unnoticed nature-affirming moral advice.[2]

After Einstein, however, we have a whole new way of looking at our ecological predicament—an unprecedented cosmological point of view. The preceding chapters in this book have dealt with a number of topics, but primarily with the connection between the idea of God and the scientifically based drama of cosmic awakening. Here I want to understand the meaning of ecological responsibility in terms of the anticipatory theological and cosmological perspective I have been advocating. This new perspective gives us, I believe, a fresh set of incentives with which to approach the present crisis. What is at stake is not just the well-being of life on our planet but, in a way, the future of the universe. If the universe is a drama of awakening, as I have proposed, then the existence and flourishing of life and other emergent outcomes on planet Earth are not just a sideshow. The future of life is a cosmic, not just a terrestrial, concern.

The Nicene Creed's central tenet is that God has entered everlastingly and irreversibly into time, into the cosmic story. The doctrine of God's incarnation means that caring for the natural world is simultaneously caring for the vulnerable God who has risked becoming fully united with irreversible time. The incarnational perspective of Christian faith should long ago have inspired Christians, of all people, to care for the natural world. The fact that until recently we have mostly ignored questions about the well-being of the nonhuman world testifies to the dominance of the analogical over the anticipatory stance in the shaping of Christian spirituality. Although analogy in principle supports an ecologically rich sacramental sense of nature, its habitual subordination of time to timelessness is ecologically problematic. Here I want to examine whether the Abrahamic anticipatory stance of biblical faith, along with our new awareness of an unfinished universe, can support the establishment of a vigorous intergenerational Christian ecological vision.

It strikes me as ironic that our ecological theologies so far have been dominantly apologetic and analogical in their theological style; that they have overlooked the anticipatory theme that the whole universe is a dramatic awakening; and that the natural world is open not only to being healed but also to being given a new and unprecedented future. When secular environmentalists complain about Christianity's indifference to the natural world, they are usually thinking about the Platonic notion of the survival of individual souls after death rather than about the central Christian teaching that God has entered irrevocably into time.[3] Their criticism is worth examining. If people are concerned solely about their personal salvation outside of time, will they not remain indifferent to the long-term welfare of the natural world?[4]

An anticipatory Christian ecological theology, I suggest, will not overlook the theme of promise for the fulfillment of creation that the majority of biblical scholars since the late nineteenth and early twentieth century have taken to be a defining characteristic of Jesus's own faith and that of his earliest followers.[5] Primitive Christianity was predominantly anticipatory rather than analogical in its worldview. The Bible, if read carefully, seeks a new future not only for the people of God but also for the whole of creation.⁶ The anticipatory discourse of the prophets—underneath their sometimes threatening language about divine judgment—was intended to keep alive the general biblical theme of confidence in a God who makes all things new, who makes and keeps promises, and who longs for the fulfillment of all creation.⁷

From the beginning, as we have seen, Christians hoped for the healing of all creation. Most Christian theologies of nature have, however, failed to allow that the universe may still have a future of new creation and fuller being up ahead.⁸ Anticipatory faith, on the other hand, passionately hopes that the coming of the "reign of God" and the sending out of the Holy Spirit will literally renew the face of the earth.

Three Ways of Reading Nature

How we treat the natural world after Einstein depends on how we read the cosmic story. Let us recall that the archaeonomic reading acknowledges that the universe is a long process but devoid of any dramatic meaning. Can archaeonomic materialism provide good reasons for its proponents to care for nature? The analogical reading, in spite of its idealizing of timelessness, discovers a sacramental meaning in nature. It looks at the fragile flowering of life on Earth as somehow revelatory of a divine beauty that abides in full glory outside of time. But can analogy appreciate the temporal depth and dramatic beauty of our still-emerging cosmos? Is it not too eager to exit the theater of nature before giving it a fair chance to blossom fully?

The anticipatory reading is most impressed that the universe is an unfinished drama. So, in times of crisis, instead of running for the exits, it undertakes a vigil of active expectation and remains open to the coming of a new future for creation. The anticipatory way of reading the cosmic story, I suggest, allows for a fresh ecological, spiritual, and ethical outlook. But before developing further the outline of an anticipatory ecological theology, let us look more closely at the archaeonomic and analogical ways of understanding the well-being of nature.

Archaeonomy. This, as we have seen, is the contemporary, academically certified, materialist way of looking at the natural world after Einstein.

The archaeonomic stance gives rise to a comprehensive worldview according to which living cells, organisms, and nervous systems are at bottom nothing more than lifeless elements now fleetingly gathered together into elaborate living syntheses. Archaeonomy takes the initially lifeless cosmic stage of primordial subatomic dispersal to be the ground state of the world's being. All presently living and thinking beings, accordingly, are reducible to mere aggregates of nonliving subatomic units that physics alone is equipped to understand. Life, according to archaeonomic materialism, is in principle nothing more than the lifeless stuff from which everything in nature is made.[9]

Archaeonomy, therefore, formally rules out the possibility that life can have any lasting value, since life (formally speaking) does not even have, nor will ever have, real existence. It is quite surprising, then, that a good many archaeoomic materialists are numbered among our most passionate environmentalists. The best example I can think of is the renowned Harvard entomologist E. O. Wilson. Wilson is a strict materialist who is obliged by his creed to reduce nature ontologically to lifeless and mindless physical units. And yet it is hard to find anyone who cares for life and nature's well-being more fervently than Wilson.[10] For this hardcore scientific materialist, any attribution of intrinsic meaning or value to life is by definition illusory. Nevertheless, as Wilson's own life exemplifies, we need to treasure and care for the life-world anyway.

But why? What is there about archaeonomic materialism that can provide a solid intellectual and spiritual foundation for human environmental responsibility? All of life, the consistent materialist has to assume, is heading toward final and absolute death. If life's destiny is bare nothingness, what is there about life that calls for an ethical ecological concern here and now? Is life precious, perhaps, simply because it is precarious?

In my opinion, Wilson does not really believe that life is reducible to nonlife any more than he believes that his mind is reducible to mindlessness. He loves life, and rightly so, but his archaeonomic worldview cannot logically support or intellectually justify his exemplary biophilia. Like most other materialists, he has an ethical sensitivity that remains secretly tied to the analogical worldview that still quietly shapes the everyday lives and ethical assumptions of most people. Analogy, perennialism, and vitalism may have lost intellectual stature today, but even in the secular culture of scientific naturalism they continue to play a considerable, if hidden, role in the shaping of ethical life.

Archaeonomists are openly critical of analogical Christianity because they believe its expectation of personal immortality in a timeless heaven beyond death weakens the ecological resolve to care for this world here and now. But can Wilson's archaeonomic worldview, which begins with the belief that lifeless matter is the basic stuff of the universe, provide better reasons to care for life on Earth? Archaeonomists, after all, are convinced that life in the Big Bang universe will eventually disappear for good. If so, perhaps a good reason to care for life lies in an implicit obligation to preserve as long as possible the fragility and delicacy of its beauty. Secular environmentalists sometimes claim that Christians who look forward to "the life of the world to come" are less likely to appreciate the evanescent quality of life on Earth than are scientifically educated people who are convinced that the natural beauty of nature is rare and perishable and that the flourishing of life on Earth deserves a special kind of protection on the part of human beings.[11]

Supposedly, then, the archaeonomic elimination of any prospect of final, otherworldly redemption for human souls frees up our moral energy to concentrate on the preciousness of life on Earth here and now, undistracted by analogical optimism. If God does not exist, life could never have been intended to exist at all. Maybe it is because life was not intended to be here that we should not take it for granted. Perhaps we can even love life all the more because of its delicacy and vulnerability in the face of constant threats to its very existence in an indifferent universe.

Of course, we also have pragmatic reasons to care for nature. For example, we humans cannot live, survive, or thrive unless the natural world is healthy enough to satisfy our appetites and provide for our needs. In the formation of our ethical lives, pragmatic motivations are always worth taking into account. For example, like other species of life, we humans need a suitable terrestrial habitat, and if we are to take care of it well, we need to feel at home on Earth and in the universe. If we suspected, along with analogical religion, that nature is not our true home, would we have sufficient incentive to take care of it? If we are just passing through the world on our way to a timeless heaven, are we not more likely to be environmentally reckless than caring?

Archaeonomic naturalists claim that nature is our only, and final, home. So what we have to worry about immediately is saving our terrestrial habitat, not our souls. The cosmos is "all that it is, all there ever was, and all there ever will be," says the scientific naturalist Carl Sagan.[12] The universe is godless; life is accidental, rare, and destined to perish; and there is nothing to look forward to after death. If so, shouldn't we devote our moral energy to improving this world instead of selfishly preparing our souls for the next—which doesn't exist anyway?

Analogy. The analogical stance implies that the rareness or precariousness of life is not a sufficient reason for ecological concern. By insisting that life is reducible to lifeless and mindless elements, archaeonomists have undermined in principle any claim that life has a special value that would make it worth caring for. Precariousness alone cannot be enough to justify the depth of care needed to save life on our planet. Consequently, according to analogy, Wilson's archaeonomic materialism cannot reasonably support his outstanding ecological sensitivity. Another worldview is needed. Isn't analogy the ecologically appropriate stance?

To most Christians, it seems that the only acceptable metaphysical alternative to materialist archaeonomy is the classical analogical vision of timeless perfection that still underlies most theological interpretations of Christian faith. For analogical theology the "really real" world is the timeless realm of divine perfection that radically transcends the natural world. This heavenly arena—our true home—is immune to all becoming and perishing. The physical universe, on the other hand, is made up of perishable things whose value corresponds not to their rarity but to the degree of their transparency to an eternal beauty not of this world. Fragile living beings on Earth have special value not because they are perishable but because for a brief time they share in, and reveal to us humans, the reality of an eternally living God. The special participation of all living beings—and not just humans—in the eternal life of God allows them to function as symbols without which we would have no sense of God at all.

The analogical reading of nature finds in the beauty of both living and nonliving beings a finite manifestation of the infinite beauty, goodness, and vitality of God's timeless being. Sunshine, oceans, rivers, waterfalls, winds, storms, breath, clean water, the hardness of rocks, the blooming of plants, the majesty of trees, the alluring dark depth of forests, the expansiveness of deserts, the fertility of soil, and (prior to the age of science) the splendor of the changeless heavens above—all of these physically impressive phenomena reveal to human souls something of the infinite beauty of God, though always imperfectly.

Reading nature analogically, therefore, has the positive ecological effect of reminding us that although nature is not God, it imparts to us something of God and allows us to partake of the being of God in an imperfect way even during our earthly sojourns. To analogically minded Christian believers this means that nature is not just material for engineering projects. Analogy's sacramental reading of nature encircles life with a halo that protects it from our human acquisitiveness.[13] Analogy values nature, not because it is precarious, but because its fragile beauties give us a glimpse of the Creator's timeless goodness and beauty. In contrast to archaeonomic materialism, analogy is prepared to accept the perishability of nature without despair because after temporal beings are gone, the eternal rightness to which they have fleetingly pointed still remains. Since analogy also looks forward to an eternal communion of the soul with God after death, it allows us to let go of terrestrial things after they have served to let in the greater light of divinity. And our being willing to let go of earthly things should keep us from squeezing the life out of our terrestrial place of exile. This quality of letting go, analogy insists, is essential to ecological responsibility.

By acknowledging the revelatory value of nature, the analogical vision, properly understood, teaches us to tread softly on planet Earth. Analogy loves nature without absolutizing it. Analogy assumes—as Saint Augustine taught—that we are all made for God, so only our final communion with God's timeless perfection can bring peace to our souls. By opening our hearts to the timelessness of God here and now, analogical faith lets us enjoy the temporal flowering of nature while also keeping our souls detached from the natural world. This detachment is not painful as long as our desires can be diverted toward the eternity of God. Analogically speaking, therefore, the premier ecological virtue is gratitude. Developing the habit of giving thanks allows us to enjoy finite beings without draining them of their inner substance. Analogy allows us to feast gratefully on the banquet of creation while restraining our tendencies to consume it completely.

When modern thought let go of God, it did not strip us of the insatiable need we have for infinity. Archaeonomy's intellectual erasure of transcendence has not eliminated the limitlessness of human longing. Consequently, in the absence of God we insatiable humans have projected our ineradicable longing for infinity idolatrously onto our fragile, finite planet—thus causing us greedily to strip it of its revelatory sacramental vitality. The analogical stance, on the other hand, allows us to embrace both the infinity of our longing and the limitations of nature.

Anticipation. An analogical reading of nature is indispensable to any ecological theology that professes to be rooted in Christian tradition. Nevertheless, the analogical reading, while rightly acknowledging the goodness of nature, fails to appreciate the dramatic fact that the universe is still coming into being. Analogy holds onto a vision of nature that flourished long before the age of science and that continues to shape spiritual and ethical life all over the planet. After Darwin and Einstein, however, we now realize that the universe is not a finished product and that its narrative development is not over. It exists as a creation still capable of more and fuller being.

Analogical spirituality may rightly continue to inform Christian ecological theology, but it fails to bring out the ecological significance of our new scientific understanding of an awakening universe. While post-Einsteinian science has been stretching our sense of time, biblical scholars have been recovering the early Christian sense that the whole of creation, not just the individual soul, anticipates a glorious new future. In the Bible, the characteristic mode of divine influence in the world is that of making creation new.[14] As the basis for a Christian theology of nature, therefore, the biblical theme of promise for the future, along with the anticipatory sense of God as somehow not-yet, may be linked up now with the new scientific picture of an unfinished universe.

One of the drawbacks of a purely analogical reading of nature is that its hope is too small. Analogical ideas about personal salvation may lead to an exaggeration of the human soul's detachment from nature during its journey into God. Because of its prescientific reading of the natural world, analogy fails to take into account the possibility that the universe itself is an ongoing dramatic journey of transformation. Instead, analogy distracts us from cosmic hope by turning our attention to an imagined perfection existing in fully polished splendor outside the natural world.

In that case, our personal religious journeys may amount to little more than a cleansing of our souls of time. Analogy ignores the prospect of our participating with enthusiasm in the long self-transformation of an awakening universe. The analogical stance assumes that everything of significance or value in the natural world has already been realized in eternity. The result of that emphasis is that the analogical reading fails to hope sufficiently for the natural world's future. In its reading of the natural world analogy risks reducing human ethical life on Earth to a spinning of our moral wheels in a squirrel cage from which, after proving our spiritual stamina, we may finally aspire to be released.[15] According to the analogical perspective, the new scientific story of an unfinished universe is therefore more of a curiosity than a starting point for a theology of nature. The cosmos, and specifically the earth, can still be too easily interpreted as an obstacle course for the believer's private spiritual journey.

An anticipatory reading of the story of nature, by comparison, looks forward to the fulfillment of all time, including all of human history. Informed by both scientific cosmology and the biblical spirit of hope, anticipation stays inside the universe that is still coming into being. Anticipatory Christian faith does not long to leave the temporal world behind by an abrupt exodus into eternity. Instead, it feels the promissory winds in Israel's formation blowing through all of nature as well. In our new cosmic setting, nature and Scripture are interwoven strands of a single anticipatory narrative. The same divine word that hovered over creation in the beginning—and that became incarnate in time—still breathes the spirit of promise into all things.

A scientifically informed ecological vision, therefore, may see God as creating the world not from out of the past but from out of the future. The Holy Spirit is the breath of a divine promise renewing the entire universe and not just human existence. God's promise to Abraham is intended not just for the "people of God" but also, as the apostle Paul came to see, for the "whole of creation."[16] An anticipatory reading of nature looks beyond a purely analogical sense of nature's perishable beauties, and beyond archaeonomic pessimism, to the full awakening of the universe in the refreshing embrace of divine beauty.

Summary

The analogical worldview is out of touch with the full reality of time and the long cosmic story. Bound as it is to a predominantly prescientific, overly spatialized understanding of the natural world, analogy cannot fully apprehend the temporal depth and dramatic suspense of a world that is still coming into being. New scientific awareness of the immensely long sequence of moments and epochs in natural history, as we now realize, has exposed the relative poverty of the vertical, static, and hierarchical pictures of the world that have served analogical piety for centuries.

An anticipatory reading of the universe, I am arguing, encourages us to look for nature's cumulative dramatic value. Neither archaeonomy nor analogy lets us feel the spectacular significance of sheer duration or to appreciate the value of what can be accomplished dramatically only over long periods of natural history. Archaeonomy, as I have noted, methodically ignores the narrative texture that gives meaning and accumulating value to nature. Archaeonomy, no less than analogy, flees from the reality and significance of time by reducing everything that has occurred in cosmic history, including the emergence of lives and minds, to the pretemporal elemental simplicity from which everything arose. Analogy, for its part, anachronistically reduces cosmic time to little more than a curious deviation from the eternal present.

An anticipatory reading of the cosmic story, as I have proposed, holds time and nature together while also allowing for hope that the cosmic process can lead to indeterminate new epochs of creation. It follows, then, that an anticipatory vision of nature requires new thoughts about the biblical God's commandment to humans to be faithful "stewards" of creation. Nature is not only sacrament but also promise. Environmental conservation alone, then, is not enough. Conservation is necessary, of course, and we humans are obliged to respect and protect the immense drama of creativity that has been occurring for billions of years prior to our own arrival. Now, however, stewardship must mean preparation, not just preservation.

An anticipatory reading of nature moves us, therefore, to care for the cosmos because, for all we know, the natural world is pregnant with an incalculable future far beyond what we humans can presently predict or plan. If gratitude is the characteristic ecological virtue in the analogical vision, then hope is the fundamental ecological virtue in the anticipatory reading of nature.[17] We cherish the natural world not only because it is revelatory of God but also because it is pregnant with the Absolute Future on which life leans as its true foundation and destiny.[18]

Conclusion

CHRISTIAN THEOLOGY IS OBLIGED to think about the universe in light of the Bible, of course, but also in light of the Nicene Creed (a translation of which is included at the end of this final chapter for easy reference). The adjective "Nicene" comes from the city of Nicaea (the present-day Turkish town of Iznik), where a famous church council, convened by Emperor Constantine, took place in the year 325 to deal with disputes about what it means to say that Jesus is the Son of God. Son of God is a title given to Jesus both in his lifetime and in the New Testament. The Nicene Creed is the outcome of considerable controversy about the identity of Jesus as the Son of God. This formulation of Christian belief underwent refinement at the First Council of Constantinople in 381, and it is this later version of the Creed that most Christian churches follow today. After Nicaea and Constantinople several other councils, especially the Council of Chalcedon in 451, attempted to define more precisely the relationship between Jesus's humanity and his divinity. These attempts at clarification have never satisfied everyone, and today the ancient credal attempts to identify Jesus remain controversial.

The ongoing disputes about the Nicene Creed point to two questions addressed by religions everywhere: Does the world make sense? And do our lives have lasting significance? The Creed responds to these universal concerns by exhorting Christians not to think about the world or their own existence apart from God, and not to think about God without thinking about the man from Nazareth named Jesus. Jesus, who came to be called the Messiah, or Christ, lived for only slightly more than three decades before being crucified as a godless criminal outside the gates of Jerusalem around two thousand years ago. Were it not for this prophetic visionary and the impression he made on a group of mostly illiterate disciples several centuries before the Council of Nicaea, the Creed would never have taken shape, and nobody today would be bothered about its meaning.

There can be no reasonable doubt that after Jesus's humiliating death something unusually enlivening and empowering happened to those who had been with him during his lifetime. Immediately following their teacher's execution, some of his disciples returned home to the hill country of Galilee, but not long afterward they felt emboldened to return to the dangerous city of Jerusalem in Judaea. There they experienced a profound sense of common life and came to agree that Jesus, whose death they had recently witnessed, was now more alive and more present to them than he had been before. His disciples confessed that Jesus had been "resurrected," and they felt that the same enlivening Spirit that had empowered Jesus during his brief lifetime was now giving them new life as well. Were it not for the palpable experience of their teacher's aliveness and healing presence soon after he died, his memory would have faded for good. There would have been no Gospels, no Christianity, and no Nicene Creed.

After Einstein, why should we still be interested in the ancient theological controversies that culminated in the Nicene Creed? The answer is simple: because they have to do with the meaning and value of time—and hence the meaning of our own lives in time. The Creed outlines a hopeful sense of time that has given comfort to many generations of humans in the face of suffering and perishing. Time is lovable to Christians because it holds the promise of fulfillment rather than final annihilation. The Nicene Creed has become Christianity's main official profession of trust that time has a permanent significance and that rightness will be victorious over wrongness. Though clothed in formal philosophical and theological language, the Creed's claims about Jesus respond to a common human longing for meaning and courage in the face of evil, the threat of transience, and the possibility of nonbeing.[1]

The Nicene Creed assumes that only what is indestructible can relieve our anxiety about death and give enduring meaning to our lives. This is why Christians believe that divine imperishability must be protected at all costs. And yet the Creed also implies that if God stayed outside of time, the world—with its perpetual perishing—would remain unhealed.

Throughout the centuries Christians have been zealous in protecting their indestructible God from getting tied up too closely with the perishable physical world, but their Creed forbids any separation of God from time. It claims that the man Jesus, the very incarnation of God, has entered fully into the stream of time. When it professes that Jesus is the Son of God and that the Son has existed "before all ages," it means not only that time is real but also that it is joined forever to what is incorruptible.

If there is any teaching that differentiates early Christianity from competing worldviews, it is the belief that time has everlasting value. In the Creed's identification of divine fidelity with the self-sacrificing love of Jesus it differs from other belief systems, most of which protect indestructibility by keeping it from getting mixed up with time. The Nicene Creed affirms that rightness dawned in time decisively in the man Jesus of Nazareth. It claims that the risen Jesus is present to the whole universe through the power of the Holy Spirit, the vitalizing breath of God that the Creed takes to be "the Lord, the giver of life." In expressing belief in "the resurrection of the dead and the life of the world to come," the Creed refuses to separate God from what goes on in time. Unlike Einstein and Spinoza, it does not abolish time but instead gives time an inexhaustible future.

The Nicene Creed also ties Christian life and faith to the existence of "one, holy, catholic, and apostolic Church." The church (*ecclesia*) is the body of those "called out" to keep hope alive *in time* for the climactic dawning of rightness in which the whole creation, "all things, visible and invisible," finally reaches everlasting fulfillment. The church, the universal (catholic) community of hope, exists to greet the whole world with the good news that in Jesus rightness has dawned in time decisively and that time therefore is not the dead end we may have thought it to be.

Underneath its formality, the Creed is an expression of Christianity's consoling attempt to keep alive, in each new age, Jesus's oft-repeated words "Be not afraid." Christians who recite the Creed today, whether ritually along with others or in the quiet of their own hearts, may now feel exalted rather than diminished by new scientific discoveries of the immense age and size of the cosmos. They are encouraged to hope that the final conquering of wrongness will not only heal their own lives but will also bring fulfillment to the whole universe, which, after Einstein, extends far beyond what previous generations of Christians could ever have imagined.

Einstein's main importance for theology, therefore, lies not in his love of eternity, nor in his ideas about religion, faith, and God, but in his success in making matter inseparable from time, which requires us to take note of the narrative complexion of the universe. Einstein himself did not seem to notice the narrative. When he talked about God as the Old One, or when he said he admired Spinoza's God, he was exalting eternity over time. He was content with a perfectly timeless geometric absolute that never had to care for or empathize with what actually happened in time. By connecting the cosmos tightly to the kingdom of pure numbers he was seeking to make nature imperishable.

Yet it is the narrative, temporal aspect of nature that allows it to be the carrier of dramatic meaning. The Nicene Creed is one of many human attempts to say what this meaning is. According to the Creed, perishing and the threat of meaninglessness are defeated by an infinite self-sacrificing love that enters into every moment of time, drawing it toward an everlasting future fulfillment—when God will be "all in all."[2] God's incarnation in Jesus, therefore, implies that everything that enters into the formation of the bodily existence of this specially revered human being is joined forever to the life of God. Since flesh is matter, and time is inseparable from matter, the "enfleshed" God is anything but pure timelessness. Human destiny, then, cannot consist of an escape from time into timelessness. Christians hope for a climactic communion of creation with God, but it is with the God whose identity is inseparable from what happens in time.

Since the existence of life and thought are narratively intertwined with the entire physical universe, in hoping for redemption we cannot contemplate leaving the universe outside the gates of paradise. Every moment of the dramatic passage of the universe in time is imprinted indelibly on the indestructible rightness that awakens it. That the cosmos has the form of a dramatic story is possible, however, only if time is real and inseparable from matter. Even though the physical universe may eventually decay energetically, the drama of awakening into which matter and time are sublimated enters everlastingly into the life-story of God, contributing in a way to the very identity of God. Relativity, in this regard, opens up the prospect that God's incarnation in matter is much deeper than theologians had previously assumed. Similarly, it implies that resurrection is a more cosmic occurrence than most previous Christian thought had earlier imagined.

If we look at the narrative and not just the geometry, we can see that the universe has always been a drama of awakening. The cosmic awakening was already underway long before the human species showed up. Before we came along, the story of cosmic awakening included the emergence of life and, before that, the three immensities of space, time, and complexity. The cosmic awakening has continued in the emergence of thought, faith, freedom, hope, compassion, and care. The universe finds its purpose, its point, its meaning, in sponsoring the awakening and flourishing of these precious outcomes. We find meaning in our own lives by participating in, preserving, and enhancing the grand cosmic drama of awakening. Our science and technology can also find meaning, especially in the digital age, by contributing novelty and extension to the cosmic adventure already under way.

The end of cosmic time is nowhere in sight. The point of human existence, therefore, must include the vocation of keeping the cosmos open to the increase of life, mind, freedom, faith, hope, and love as we await the future. For humans, living a good life means contributing to the drama of cosmic awakening. And if unknown universes exist beyond our own, an anticipatory theology offers a hope wide enough to bring them, too, into communion with the emerging beauty already latent in the divine creation of "all things, visible and invisible."

Becoming aware scientifically of what is going on in the universe, therefore, is not irrelevant to answering the Kantian question of what we should be doing with our lives. After Einstein our human obligation in this universe is, at the very least, to foster, both now and through future generations, hope for the increase of cosmic awakening that began long before we humans appeared in the story. For Christians, this vocation is realized concretely in the imitation of Christ.

If hope is wide enough, it also seeks a future for whatever and whoever has already perished. "Hope that is seen is not hope," says the apostle

Paul.³ In the face of perishing and death, hope is hard to keep alive, since we are eager to see but reluctant to wait. But even waiting may be filled with meaning as long as we are attentive to the opportunities that time provides for acts of love. And love is liberated most effectively when we can see to it that those we love have a future. Consequently, the good life includes looking for reasons to hope and then passing these on to others, including our descendants.

We would have no serious incentive to do so if we surrendered our minds and souls exclusively to either the archaeonomic or the analogical stance. Each of these is without hope for the future of the universe. Being without cosmic hope, neither archaeonomy nor analogy is willing to wait for and work toward the fuller awakening of the universe. As we have seen, they are both too impatient to savor a universe that is still aborning. Only anticipation, as I see it, embraces the challenge of taking the long view—which includes hoping for the final abolition of suffering and death.

Both taking a detour into analogical optimism and resting archaeonomically on the fixity of the cosmic past are tempting—and I believe understandable—ways to use up the time given to each human life. The universe, however, has been waiting actively and fruitfully for many billions of years. The wait has been long, but given the amazing outcomes so far, it has not been a waste of time. Furthermore, waiting in real time, no matter how long it lasts, can be filled even now with a special kind of happiness. In an unfinished universe anticipation *is* the happiness of the present.⁴ Our waiting with active anticipation serves life because we are resisting the "sweet decay" of giving up in despair.⁵ By embracing the vitalizing disposition of striving and anticipating, humans can contribute even now to the divine cause of liberating life and awakening the universe.

The Nicene Creed
    I believe in one God,
        the Father almighty,
    maker of heaven and earth,
    of all things, visible and invisible.
    I believe in one Lord Jesus Christ,

the Only Begotten Son of God,
born of the Father before all ages.
God from God, Light from Light,
true God from true God
begotten, not made, consubstantial with the Father;
through him all things were made.
For us men and for our salvation
he came down from heaven,
and by the Holy Spirit was incarnate of the Virgin Mary,
and became man.
For our sake he was crucified under Pontius Pilate,
he suffered death and was buried,
and rose again on the third day
in accordance with the Scriptures.
He ascended into heaven
and is seated at the right hand of the Father.
He will come again in glory
to judge the living and the dead
and his kingdom will have no end.
I believe in the Holy Spirit, the Lord, the giver of life,
who proceeds from the Father and the Son,
who with the Father and the Son is adored and glorified,
who has spoken through the prophets.
I believe in one, holy, catholic and apostolic Church
I confess one Baptism for the forgiveness of sins
and I look forward to the resurrection of the dead
and the life of the world to come. Amen.

Notes

Introduction

1. In the conclusion to this book, I give the full text of the Nicene Creed with a brief explanation of where it came from, what it means to Christians today, and what it looks like theologically after Einstein.
2. John F. Haught, *God after Darwin: A Theology of Evolution,* 2nd edition (Boulder. Colo.: Westview, 2008).
3. The title is also appropriate because throughout I shall be discussing Einstein's own ideas of God.
4. I do not duplicate, for example, Max Jammer's discussion of the theological implications of Einstein's physics. Max Jammer, *Einstein and Religion: Physics and Theology,* (Princeton, N.J.: Princeton University Press, 1999), 155–265.
5. See, for example, the discussion by Paul Davies in his *God and the New Physics* (New York: Simon and Schuster, 1994).
6. A good example of this "concordist" way of reading the Bible is Gerald Schroeder's *Genesis and the Big Bang: The Discovery of Harmony between Modern Science and the Bible* (New York: Bantam, 1990).
7. See, for example, E. Stump and N. Kretzman, "Eternity," *Journal of Philosophy* 78 (1981), 429–452.

Chapter 1. God

Epigraph 1: Albert Einstein, *Ideas and Opinions,* trans. Sonja Bargmann (New York: Crown, 1994), 48.
Epigraph 2: Ibid., 47.
Epigraph 3: Quotations from the Nicene Creed throughout this book are from the official American English Roman Catholic liturgy, https://www.usccb.org/committees/divine-worship/policies/mass-settings-texts#tab—nicene-creed. Not all quotes from the Creed are identified as such.

1. As stated in the Nicene Creed.
2. See Pierre Teilhard de Chardin, *The Human Phenomenon,* trans. Sarah Appleton-Weber (Portland, Ore.: Sussex Academic, 1999).
3. Here "rightness" means not merely moral rightness but also intellectual, metaphysical, and aesthetic rightness. For a book-length development of the notion that religion is an awakening to indestructible rightness see John F.

Haught, *The New Cosmic Story: Inside Our Awakening Universe* (New Haven: Yale University Press, 2017).

4. The philosopher Karl Jaspers (1883–1969) referred to the historical period during which our main religious and philosophical traditions arose (roughly between 800 and 300 BCE) as the "axial age." It was during this period, he says, that "the spiritual foundations of humanity were laid simultaneously and independently in China, India, Persia, Judea, and Greece." Karl Jaspers, *The Way to Wisdom: An Introduction to Philosophy* (New Haven: Yale University Press, 1951), 98.

5. Micah 6:8. Scriptural quotations in this book, unless indicated otherwise, are from the New Revised Standard Version (NRSV).

6. See, for example, Luke 1:46–55, 67–79.

7. Pierre Teilhard de Chardin, *Christianity and Evolution,* trans. René Hague (New York: Harvest, 1974), 240.

8. This understanding of divine compassion has been expressed especially well in "process theology." Here I am indebted in part to aspects of the process perspective as laid out by Alfred North Whitehead. See his *Process and Reality,* corrected ed., ed. David Ray Griffin and Donald W. Sherburne (New York: Free Press, 1968), 29, 34–51, 60, 81–82, 86–104, 340–351; and Alfred North Whitehead, "Immortality," in *The Philosophy of Alfred North Whitehead,* ed. Paul A. Schilpp (Evanston, Ill.: Northwestern University Press, 1941), 682–700. See also Charles Hartshorne, *The Logic of Perfection* (Lasalle, Ill.: Open Court, 1962), 24–62, 250; and John B. Cobb Jr. and David Griffin, *Process Theology: An Introductory Exposition* (Philadelphia: Westminster, 1976). Although I do not embrace every aspect of process thought, I accept Whitehead's understanding of the world as made up of events that, even while perishing, remain present and "objectively immortal" in the compassionate experience of God. I also want to make room for personal or subjective survival of death, though not independently of the survival of the entire cosmic story within the compassionate love of God.

9. John Macquarrie, *The Humility of God* (Philadelphia: Westminster, 1978), 34.

10. Pope John Paul II, *Encyclical Letter Fides et Ratio of the Supreme Pontiff to the Bishops of the Catholic Church on the Relationship between Faith and Reason* (1998), #93. http://www.vatican.va/content/john-paul-ii/en/encyclicals/documents/hf_jp-ii_enc_14091998_fides-et-ratio.html.

11. Jürgen Moltmann, *The Crucified God,* trans R. A. Wilson and John Bowden (New York: Harper and Row, 1973), 195.

12. In *The New Cosmic Story: Inside Our Awakening Universe* (New Haven: Yale

University Press, 2017), I write about religion and cosmology in a more general way than in this book. Here I want to bring out what is distinctively Christian in our new understanding of a cosmos awakening to indestructible rightness.

13. *The Ante-Nicene Fathers,* ed. and trans. Alexander Roberts and James Donaldson, vol. 1 (Buffalo, N.Y.: Christian Literature, 1884–1886), 352, 459, 414–416, 447–449, cited in William C. Placher and Derek R. Nelson, eds., *Readings in the History of Christian Theology: From Its Beginnings to the Eve of the Reformation,* vol. 1, rev. ed. (Louisville, Ky.: Westminster/John Knox Press, 2015).

14. "One in being with the Father" is an alternative translation of "consubstantial with the Father." A full recent translation and a more extended summary of the Nicene Creed and its implications for a theological understanding of time is provided in this book's concluding chapter.

15. Belonging or being open to the future is in no sense a way of escaping the present. To exist fully in the present requires an appreciation of the horizon of what is not-yet. If we fail to keep our eyes on the future, we end up absolutizing the present by expecting too much of it. See Jürgen Moltmann, *Theology of Hope: On the Ground and the Implications of a Christian Eschatology,* trans. James W. Leitch (New York: Harper and Row, 1967), 32. Moltmann asks whether hope for the future can cheat us of the happiness of the present. He answers: "How could it do so! For [hope] is itself the happiness of the present." Ibid.

16. The idea of God as the Absolute Future is developed by the Catholic theologian Karl Rahner, in *Theological Investigations,* vol. 6, trans. Karl and Boniface Kruger (Baltimore, Md.: Helicon, 1969), 59–68.

17. Lee Smolin, *Time Reborn: From the Crisis in Physics to the Future of the Universe* (New York: Houghton Mifflin Harcourt, 2013), xiii.

Chapter 2. Eternity

Epigraph 1: Albert Einstein, quoted in the *New York Times,* April 25, 1929, cited in *The Ultimate Quotable Einstein,* ed. Alice Calaprice (Princeton, N.J.: Princeton University Press, 2011), 336.

Epigraph 2: Albert Einstein, *Ideas and Opinions,* trans. Sonja Bargmann (New York: Crown, 1994), 261–262.

1. For details see the fine book by Jimena Canales, *The Physicist and the Philosopher: Einstein, Bergson, and the Debate That Changed Our Understanding of Time* (Princeton, N.J.: Princeton University Press, 2015), 197–199.

2. Helen Dukas and Banesh Hoffman, eds., *Albert Einstein, the Human Side:*

Glimpses from His Archives (Princeton, N.J.: Princeton University Press, 2013), 32–33.
3. Cited in Denis Brian, *Einstein: A Life* (New York: Wiley, 1996), 161.
4. Albert Einstein, *Ideas and Opinions*, 38–40.
5. The New Atheist Sam Harris, for example, defines faith of all sorts as "belief without evidence." Harris, *The End of Faith: Religion, Terror, and the Future of Reason* (New York: Norton, 2004), 58–73, 85.
6. Quoted in Banesh Hoffmann, *Albert Einstein: Creator and Rebel* (New York: Viking, 1972).
7. *Albert Einstein—Michele Besso Correspondence, 1903–1955* (Paris: Hermann, 1972), 537–538, cited in Max Jammer, *Einstein and Religion* (Princeton, N.J.: Princeton University Press), 161.
8. Like Paul Tillich, I doubt that there are many pure pantheists. Here I have been listing the logical implications of pantheism for our understanding of the universe. Psychologically, people who refer to themselves as pantheists usually deviate in one way or another from the formal definition of the term. See Paul Tillich, *Systematic Theology*, vol. 1 (Chicago: University of Chicago Press, 1963), 233–237. On the topic of Einstein and freedom, see Chapter 9 below.
9. Whether the Big Bang universe really had a beginning in time and what this could possibly mean both scientifically and theologically are questions taken up in Chapter 6 below.
10. Contemporary cosmology has, however, revived the idea of a cosmological constant, but not in such a way as to abolish irreversible time. See Chapter 6 below.
11. "Relativity," Einstein insisted, "is a purely scientific matter and has nothing to do with religion." Quoted in Philipp Frank, *Einstein: His Life and Times* (New York: Knopf, 1947), 190.
12. Michael J. Buckley, *At the Origins of Modern Atheism* (New Haven: Yale University Press, 1990). Buckley argues that the intellectual roots of contemporary atheism lie in grounding the idea of God in geometry rather than in religious experience.
13. John F. Haught, *God after Darwin: A Theology of Evolution*, 2nd ed. (Boulder, Colo.: Westview), 1–11.
14. The "narrative" cosmological principle, as I call it, is distinct from the "Copernican" and "anthropic" cosmological principles, both of which are, in my view, too narrowly terrestrial in scope to capture all that is going on in the universe. A narrative cosmological principle requires that we have to wait to

find out what happens in the end, but it does not rule out that there may be good reasons for hope in the meantime.

Chapter 3. Time

Epigraph 1: Hermann Minkowski, "Space and Time," in Hendrik A. Lorentz, Albert Einstein, Hermann Minkowski, and Hermann Weyl, *The Principle of Relativity: A Collection of Original Memoirs on the Special and General Theory of Relativity* (New York: Dover, 1952), 75, cited in "Hermann Minkowski — the Man Who Posed the Problem," International Society for the Advanced Study of Spacetime, http://www.spacetimesociety.org/minkowski.html. Lee Smolin writes: "Minkowski's invention of what we now call Minkowski spacetime was a decisive step to the elimination of time, because it persuasively established that all talk of motion in time could be translated into mathematical theorems about a timeless geometry." Lee Smolin, *Time Reborn: From the Crisis in Physics to the Future of the Universe* (New York: Houghton Mifflin Harcourt, 2013), 60.

Epigraph 2: Alfred North Whitehead, *Modes of Thought* (New York: Free Press, 1968), 152.

Epigraph 3: Smolin, *Time Reborn*, xi.

1. Ibid., 25–36 and passim. Smolin himself insists on the reality and irreversibility of time. I agree with him that our experience of the movement of time from past to future is no illusion. If I read him correctly, however, for Smolin the future seems empty, whereas I believe the future, that which is not-yet, is a rich reservoir of yet unrealized possibilities. The existence of a sphere of possibilities waiting to be realized is a very good reason to hope.

2. See, for example, Max Tegmark, *Our Mathematical Universe: My Quest for the Ultimate Nature of Reality* (New York: Knopf, 2014).

3. Carlo Rovelli, *The Order of Time* (New York: Riverhead Books, 2018), 24, 161.

4. The philosopher Martin Heidegger (like Henri Bergson and Alfred North Whitehead), as Jimena Canales points out, criticized Einstein's interpretation of relativity for making the passage of time a psychic illusion: "The theory's limited notion of time," Heidegger said, "was the result of how it spatialized time." Quoted in Jimena Canales, *The Physicist and the Philosopher: Einstein, Bergson, and the Debate That Changed Our Understanding of Time* (Princeton, N. J.: Princeton University Press, 2015), 141. Whitehead would have considered Einstein's view of time to be an example of the "fallacy of

misplaced concreteness," the confusion of abstractions with concrete reality. The inability to distinguish past from future, therefore, is the result of a mistake in logic. Alfred North Whitehead, *Science and the Modern World* (New York: Free Press), 51.

5. Katie Mack, "Freeman Dyson's Quest for Eternal Life," *New York Times*, March 3, 2020. Many other scientists, in agreement with Mack, consider the irreversible trajectory of time to be bad news. In these pages I argue that this is not necessarily so. See also Mack's recent book *The End of Everything (Astrophysically Speaking)*, illus. ed. (New York: Scribner, 2020).

6. Lee Smolin, however, believes that the future is open to the evolution of not only life but also the laws of nature. Maybe so, but my point here is that the rules and regulations of nature—even if only relatively stable—are necessary conditions for, and not obstacles to, the emergence of novelty and eventual freedom in cosmic process. See Smolin, *Time Reborn*, 123–137.

7. I use the singular word "meaning" instead of the plural "meanings" because the definition of meaning, as I see it, entails the gathering of the world's plurality into an overarching unity. There is no meaning or coherence apart from a principle of unity that ties everything together. What, then, we may ask, is the universe's principle of unity, and is it discoverable?

8. As the novelist Karen Blixen (Isak Denison) has said: "All sorrows can be borne if you put them into a story or tell a story about them." https://en.wikiquote.org/wiki/Karen_Blixen.

9. A good example of an atheistic interpretation of deep time is Richard Dawkins's *Climbing Mount Improbable* (New York: Norton, 1996).

10. For the sake of convenience, I use the term "geometric" as a stand-in for all applicable varieties of mathematical representation.

11. The physicist John Archibald Wheeler (1911–2008) is often given credit for saying that "time is nature's way to keep everything from happening at once." However, Einstein also is sometimes said to be its originator. See the discussion at https://en.wikiquote.org/wiki/John_Archibald_Wheeler.

12. Proponents of intelligent design include Stephen C. Meyer, *Darwin's Doubt: The Explosive Origin of Animal Life and the Case for Intelligent Design* (New York: HarperCollins, 2014); Phillip E. Johnson, *The Wedge of Truth: Splitting the Foundations of Naturalism* (Downers Grove, Ill.: InterVarsity, 1999); Jonathan Wells, *Icons of Evolution: Science or Myth? Why Much of What We Teach About Evolution Is Wrong* (Washington, D.C.: Regnery, 2000); Michael J. Behe, *Darwin's Black Box: The Biochemical Challenge to Evolution* (New York: Free Press, 1996); and William A. Dembski, *Intelligent Design:*

The Bridge between Science and Theology (Downers Grove, Ill.: InterVarsity, 1999).

13. See Paul Tillich's profound meditation on "Waiting" in *The Shaking of the Foundations* (New York: Scribner, 1948), 149–152.
14. Pierre Teilhard de Chardin, *Activation of Energy,* trans. René Hague (New York: Harcourt Brace Jovanovich, 1970), 239.
15. My referring to God as "up ahead" and *not-yet* has been influenced especially by some works by the Jesuit geologist Pierre Teilhard de Chardin: *Writings in Time of War,* trans. René Hague (New York: Harper and Row, 1968); *The Heart of Matter,* trans. René Hague (New York: Harvest, 2002); *The Divine Milieu* (New York: Harper and Row, 1962); *Human Energy,* trans. J. M. Cohen (New York: Harvest/Harcourt Brace Jovanovich, 1962); *The Future of Man,* trans. Norman Denny (New York: Harper and Row, 1964); *How I Believe,* trans. René Hague (New York: Harper and Row, 1969); and *The Human Phenomenon,* trans. Sarah Appleton-Weber (Portland, Ore.: Sussex Academic, 1999). The meaning of what Teilhard calls consistence is developed in *The Heart of Matter,* 15–79.

Chapter 4. Mystery

Epigraph 1: Albert Einstein, "What I Believe," *Forum and Century,* 84 (1930), 193–194, cited in *The Ultimate Quotable Einstein,* ed. Alice Calaparice (Princeton, N.J.: Princeton University Press 2011), 331.

Epigraph 2: Albert Einstein, "Science, Philosophy and Religion, A Symposium," *The Conference on Science, Philosophy and Religion in Their Relation to the Democratic Way of Life, Inc.* (New York, 1941), http://www.update.uu.se/~fbendz/library/ae_scire.htm.

1. For example, B. F. Skinner, *Beyond Freedom and Dignity* (New York: Bantam, 1972), 54.
2. Albert Einstein, *Ideas and Opinions,* trans. Sonja Bargmann (New York: Crown, 1994), 11. The first epigraph to this chapter offers a somewhat different wording.
3. "The eternal mystery of the world is its comprehensibility. . . . The fact that it is comprehensible is a miracle." Einstein, *Ideas and Opinions,* 292. Often this comment is phrased as follows: "The most incomprehensible thing about the world is that it is comprehensible."
4. For the distinction between problem and mystery see Gabriel Marcel, *Being and Having,* trans. Katherine Farrer (Westminster, Md.: Dacre, 1949).

5. See Elizabeth Johnson, *Quest for the Living God: Mapping Frontiers in the Theology of God* (New York: Continuum, 2007), 25–48.
6. To repeat, the notion of God as Absolute Future is that of Karl Rahner, S.J., in *Theological Investigations,* vol. 6, trans. Karl Kruger and Boniface Kruger (Baltimore, Md.: Helicon, 1969), 59–68. See also Wolfhart Pannenberg, *Faith and Reality,* trans. John Maxwell (Philadelphia: Westminster, 1977); Pannenberg, *Toward a Theology of Nature,* ed. Ted Peters (Louisville, Ky.: Westminster/John Knox Press, 1993); and Ted Peters, *God—The World's Future: Systematic Theology for a New Era,* 2nd ed. (Minneapolis: Fortress, 2000).
7. Einstein, *Ideas and Opinions,* 11. In 2008, a letter of Einstein's was auctioned off and sold for £207,000 ($404,000). A translation of part of this letter by Alberto Martínez includes these words: "The word God is for me nothing more than the expression and product of human weaknesses, the Bible a collection of honorable but still primitive legends aplenty. No interpretation, no matter how subtle, can change this (for me)." Alberto A. Martinez, "Was Einstein Really Religious?" *Not Even Past,* April 11, 2012, https://notevenpast.org/was-einstein-really-religious-0/.
8. See Paul Tillich's response to Einstein: *Theology of Culture,* ed. Robert C. Kimball (New York: Oxford University Press, 1959), 131–132. Without denying that God is inexhaustible mystery, Tillich makes the point that mystery is "super-personal." But to inspire worship it must be at least personal. Citing the philosopher F. W. J. von Schelling, Tillich observes that only a person can heal another person. So an impersonal deity cannot heal the wounds of personal beings. See also Martin Buber, *I and Thou,* trans. Walter Kaufmann (New York: Charles Scribner's Sons, 1970).
9. Heinz Pagels, *Perfect Symmetry* (New York: Bantam, 1985), xv.
10. Psalm 18:19 (my paraphrase).
11. According to some recent estimates, our galaxy may be twice as wide as previously thought—perhaps 200,000 light-years across. Mara Johnson-Groh. "The Milky Way Just Got Larger," *Discover Magazine,* June 14, 2018, https://www.discovermagazine.com/the-sciences/the-milky-way-just-got-larger.
12. Perhaps millions of potential life-bearing planets exist in the Milky Way alone. See Dennis Overbye, "Looking for Another Earth? Here Are 300 Million, Maybe," *New York Times,* November 5, 2020, https://www.nytimes.com/2020/11/05/science/astronomy-exoplanets-kepler.html
13. For example, Matthew 6:30; 8:26.
14. Amos 5:24.
15. Matthew 5:45.
16. Here and in the following chapters when I refer to equivalence and super-

abundance, I am following the philosopher Paul Ricoeur's terminology as developed in "The Logic of Jesus, the Logic of God," in *Figuring the Sacred: Religion, Narrative, and Imagination,* ed. Mark I. Wallace, trans. David Pellauer (Minneapolis: Fortress, 1995), 279–283.
17. Matthew 20:1–16.
18. Matthew 19:14.
19. Matthew 5:38–41, 43–44. These and most other biblical references in this chapter are used by Paul Ricoeur, in "The Logic of Jesus, the Logic of God."
20. Ephesians 2:1; Colossians 2:13; Luke 15:11–32.
21. See John 9:1–12.
22. Ricoeur, "The Logic of Jesus, the Logic of God." Ricoeur admits his indebtedness for much of this interpretation to the biblical scholar Robert Tannehill's book *The Sword of His Mouth* (Missoula, Mont.: Scholar's Press, 1975).
23. Daniel C. Dennett, *Darwin's Dangerous Idea: Evolution and the Meaning of Life* (New York: Simon and Schuster, 1995). After Darwin, however, some prominent evolutionists have emphasized the role of chance or contingency in shaping the evolutionary trajectory. This is a significant development simply because it allows us to understand nature narratively and dramatically rather than mechanistically. See especially Stephen Jay Gould, *Wonderful Life: The Burgess Shale and the Nature of History* (New York: Norton, 1990).
24. Matthew 19:26.

Chapter 5. Meaning

Epigraph 1: Albert Einstein, *Ideas and Opinions,* trans. Sonja Bargmann (New York: Crown, 1994), 40.
Epigraph 2: Ludwig Wittgenstein, *Tractatus Logico-Philosophicus* (London: Routledge and Kegan Paul, 1922), 6.371.
Epigraph 3: Rainer Maria Rilke, *Letters to a Young Poet,* trans. Reginald Snell (Mineola, N.Y.: Dover, 2002), 36.
1. Einstein, *Ideas and Opinions,* 11.
2. Ibid.
3. See Pierre Teilhard de Chardin, *Activation of Energy,* trans. René Hague (New York: Harcourt Brace Jovanovich, 1970), 130–139.
4. P. W. Atkins, *The 2nd Law: Energy, Chaos, and Form* (New York: Scientific American Books, 1994), 200.
5. Jonathan Edwards, "Images of Divine Things," in *Typological Writings: The Works of Jonathan Edwards,* vol. 11, ed. W. E. Anderson, M. I. Lowance, and D. H. Watters (New Haven: Yale University Press, 1993), 62.

6. The philosopher Henri Bergson also rejects the idea that the cosmos is planned from the beginning: "A plan is a term assigned to a labor: it closes the future whose form it indicates. Before the evolution of life, on the contrary, the portals of the future remain wide open." Henri Bergson, *Creative Evolution,* trans. Arthur Mitchell (New York: Holt, 1911), 105.
7. Genesis 12–25.
8. The quotations are from John Henry Newman's famous hymn "Lead Kindly Light."
9. Isaiah 43:19; Revelation 21:5.
10. For a lengthier development of the typology outlined in this chapter see my book *The New Cosmic Story: Inside Our Awakening Universe* (New Haven: Yale University Press, 2017).

Chapter 6. Origins

Epigraph 1: Albert Einstein, *Ideas and Opinions,* trans. Sonja Bargmann (New York: Crown, 1994), 39.

Epigraph 2: Pierre Teilhard de Chardin, *Hymn of the Universe,* trans. Gerald Vann (New York: Harper Colophon, 1969), 77.

1. See William Lane Craig, *The Kalam Cosmological Argument* (Eugene, Ore.: Wipf and Stock, 2000).
2. Max Jammer echoes what I suggested earlier—namely, that "the motivation for introducing the cosmological constant may well have been the direct result of Spinoza's influence." According to George Gamow, cited by Jammer, Einstein himself eventually expressed regret for adjusting his equations to allow for his cosmological constant. Max Jammer, *Einstein and Religion: Physics and Theology* (Princeton, N.J.: Princeton University Press, 1999), 148.
3. Others are not so sure. See, for example, Ethan Siegel, "No, the Universe Cannot Be a Billion Years Younger Than We Think," *Forbes Magazine,* June 12, 2019, https://www.forbes.com/sites/startswithabang/2019/06/12/no-the-universe-cannot-be-a-billion-years-younger-than-we-think/#5e0dea9a1b6b.
4. For the meaning of "nonbeing" see Paul Tillich, *The Courage to Be* (New Haven: Yale University Press, 1952), 32–34.
5. Mircea Eliade, *Myth and Reality* (New York: Harper and Row, 1963).
6. The metaphor "ground" is used in reference to God as the "ground of being" by the theologian Paul Tillich—for example, in *The Courage to Be,* 156–186.
7. See Eliade, *Myth and Reality,* 18, 31.
8. Alex Rosenberg, writes, for example: "Everything in the universe is made up of the stuff that physics tells us fills up space." According to this Duke Uni-

versity professor of philosophy, physics "can tell us how everything in the universe works, in principle and in practice, better than anything else." He goes on: "All the processes in the universe, from atomic to bodily to mental, are purely physical processes involving fermions and bosons interacting with one another." Why should we allow particle physics to shape our worldview? "Well, it's simple, really," Rosenberg replies, "We trust science as the only way to acquire knowledge," and that, he concludes, "is why we are so confident about atheism." Rosenberg, *The Atheist's Guide to Reality: Enjoying Life without Illusions* (New York: Norton, 2012), 20–21.
9. Algernon Charles Swinburne, "The Garden of Persephone."
10. Sean Carroll, "Does the Universe Need God?" in James B. Stump and Alan G. Padgett, eds., *The Blackwell Companion to Science and Christianity* (Malden, Mass.: Wiley-Blackwell, 2012), 196. See also Sean Carroll, *The Particle at the End of the Universe: How the Hunt for the Higgs Boson Leads Us to the Edge of a New World* (New York: Penguin, 2012), 22–26.
11. See, for example, Martin Rees, *Our Cosmic Habitat* (Princeton, N.J.: Princeton University Press, 2001).
12. Pierre Teilhard de Chardin, *Writings in Time of War,* trans. René Hague (New York: Harper and Row, 1968), 81–82. At the Second Vatican Council, however, a number of the theologians who helped shape its new documents had read some of Teilhard's writings and agreed with him. The council attempted, though with only limited success, to make a transition from a one-sided analogical approach to a more anticipatory understanding of Christian faith. I discuss Teilhard's thought at greater length in my book *The Cosmic Vision of Teilhard de Chardin* (Maryknoll, N.Y.: Orbis, 2021).
13. See the poem "Spring" by Gerard Manley Hopkins, at poets.org, https://poets.org/poem/spring.
14. For theologian Paul Tillich our wonder that anything exists at all may be called "ontological shock." Tillich, *Systematic Theology*, vol. 1 (Chicago: University of Chicago Press, 1963), 113.

Chapter 7. Life

Epigraph 1: Alfred North Whitehead, *Modes of Thought* (New York: Free Press, 1968), 135.
1. Pierre Teilhard de Chardin, *Man's Place in Nature,* trans. René Hague (New York: Harper and Row, 1956), 22–24. It is to Teilhard that I owe many of my ideas on immensity and complexity, but I am not using the term "immensity" in exactly the same way as he does. In the present book the three immensities

are those of space, time, and complexity. I should add that cerebral "complexity" is measured in proportion to brain size.
2. Teilhard predicts this planetary complexity in his notion of a noosphere. Ibid., 7, 22, 79–121.
3. Lee Smolin, *Time Reborn: From the Crisis in Physics to the Future of the Universe* (New York: Houghton Mifflin Harcourt, 2013), 194.
4. Jacques Monod, *Chance and Necessity: An Essay on the Natural Philosophy of Modern Biology,* trans. Austryn Wainhouse (New York: Vintage, 1972). Monod's book is now a classic example of the archaeonomic vision of nature and life.
5. The mathematical physicist Freeman Dyson (1923–2020) even suggests that the universe seems to have known that beings endowed with consciousness were coming. Freeman Dyson, *Disturbing the Universe* (New York: Harper and Row, 1979), 250.
6. Michael Polanyi, *Personal Knowledge* (New York: Harper Torchbooks, 1964), 327ff; Polanyi, *The Tacit Dimension* (Garden City, N.Y.: Doubleday Anchor, 1967); and Polanyi, "Life's Irreducible Structure," in *Knowing and Being,* ed. Marjorie Grene (Chicago: University of Chicago Press 1969), 225–239.
7. The adjective "conative" comes from the Latin verb "to strive." As we shall see in the following chapter, the transition from conative to cognitive existence introduces a whole new epoch into our cosmic drama.
8. Polanyi, *Personal Knowledge,* 327–346.
9. Francis H. C. Crick, *Of Molecules and Men* (Seattle: University of Washington Press, 1966), 10.
10. J. D. Watson, *The Molecular Biology of the Gene* (New York: W. A. Benjamin, 1965), 67.
11. As examples of a purely archaeonomic reading of the life story, consider Richard Dawkins, *The Blind Watchmaker: Why the Evidence of Evolution Reveals a Universe without Design,* reissue edition (New York: Norton, 2015); and Daniel C. Dennett, *Darwin's Dangerous Idea: Evolution and the Meaning of Life* (New York: Simon and Schuster, 1996).
12. Whatever private feelings archaeonomists may have about life, their public worldview, as Hans Jonas has put it, is an "ontology of death." Hans Jonas, *The Phenomenon of Life* (New York; Harper and Row, 1966), 9–10.
13. See, for example, Huston Smith, *Forgotten Truth: The Common Vision of the World's Religions* (San Francisco: HarperOne, 1992).
14. This is a point made often by Pierre Teilhard de Chardin—for example, in *Activation of Energy,* trans. René Hague (New York: Harcourt Brace Jovanovich, 1970), 99–139.

15. Humans may discover in the greenness of nature, for example, a connection to the sweetness of Eden. By celebrating life here and now, in the analogical interpretation, humans taste momentarily the aliveness of creation before it was soured by sin. See Gerard Manley Hopkins's poem "God's Grandeur."
16. Henri Bergson, *Creative Evolution,* trans. Arthur Mitchell (1911; New York: Dover, 1998).
17. See Whitehead, *Modes of Thought,* 127–169.
18. Furthermore, vitalism often remains implicitly materialist or mechanistic in its understanding of the inanimate natural world. See Alfred North Whitehead, *Science and the Modern World* (New York: Free Press, 1925), 79.
19. Pierre Teilhard de Chardin, *Writings in Time of War,* trans. René Hague (New York: Harper and Row, 1967), 157–158.
20. See, for example, Martin Rees, *Just Six Numbers: The Deep Forces That Shape the Universe* (New York: Basic, 2000); and Rees, *Our Cosmic Habitat* (Princeton, N.J.: Princeton University Press, 2001).

Chapter 8. Thought

Epigraph 1: Albert Einstein, *Ideas and Opinions,* trans. Sonja Bargmann (New York: Crown, 1994), 49.

Epigraph 2: Albert Einstein, *Letters to Solvine,* Einstein Archives, 21–474, cited in *The Ultimate Quotable Einstein,* ed. Alice Calaprice (Princeton, N.J.: Princeton University Press 2011), 340.

Epigraph 3: Albert Einstein, Letter to Alfredo Rocco, November 16, 1931, cited in Calaprice, *The Ultimate Quotable Einstein,* 452.

1. For details see John Farrell, *The Day without Yesterday: Lemaître, Einstein and the Birth of Modern Cosmology* (New York: Basic, 2010). Even though Lemaître was a Roman Catholic priest, he did not read any immediate theological significance into the new cosmology made possible by Einstein's science.
2. Pierre Teilhard de Chardin, *The Human Phenomenon,* trans. Sarah Appleton-Weber (Portland, Ore.: Sussex Academic, 1999); see also C. E. Raven, *Teilhard de Chardin: Scientist and Seer* (London: Collins, 1962), 132–133.
3. Daniel C. Dennett, *Consciousness Explained* (New York: Little, Brown, 1991), 33. For my own understanding of "thought" in this chapter I am indebted in part to Bernard Lonergan, *Insight: A Study of Human Understanding* (1957; Toronto: University of Toronto Press, 1997).
4. Owen Flanagan, *The Problem of the Soul: Two Visions of Mind and How to Reconcile Them* (New York: Basic, 2002), 167–168.

5. David Sloan Wilson, *Darwin's Cathedral: Evolution, Religion, and the Nature of Society* (Chicago: University of Chicago Press, 2002), 228.
6. See Thomas Nagel, *Mind and Cosmos: Why the Materialist Neo-Darwinian Conception of Nature Is Almost Certainly False* (New York: Oxford University Press, 2012).
7. Steven Weinberg, *The First Three Minutes* (New York: Basic, 1977), 144.
8. Steven Weinberg, *Dreams of a Final Theory* (New York: Pantheon, 1992), 241–261.
9. In reaching this conclusion I have been influenced partly by the scientist and philosopher Michael Polanyi. See his books *Knowing and Being*, ed. Marjorie Grene (Chicago: University of Chicago Press, 1969), and *The Tacit Dimension* (Garden City, N.Y.: Doubleday Anchor, 1967), 31–34.
10. See also Chapter 11 below. Let me note here again that the gradual loss of heat energy is the most obvious physical reason for the irreversibility of time. For that reason, thermodynamics and not just relativity is significant theologically.

Chapter 9. Freedom

Epigraph 1: Albert Einstein, *Ideas and Opinions*, trans. Sonja Bargmann (New York: Crown, 1994), 9.

Epigraph 2: Albert Einstein, "On Freedom and Science," cited in David E. Rowe and Robert Shulmann, *Einstein on Politics: His Private Thoughts and Public Stands on Nationalism, Zionism, War, Peace, and the Bomb* (Princeton, N.J.: Princeton University Press, 2007), 435. Einstein goes on to say that freedom "is a condition in which every individual acts in accordance with his personal wishes and decisions without restriction by others. Ibid., 436.

1. Quoted in *The Ultimate Quotable Einstein*, ed. Alice Calaprice (Princeton, N.J.: Princeton University Press 2011), 332.
2. Suggestions that quantum physics and the principle of indeterminacy provide a foundation for free will are inconclusive and, I believe, illogical. Quantum indeterminacy at the micro level does not translate into freedom at the macro level. Einstein refused to accept indeterminacy even at the micro level.
3. Freedom in this second sense is championed by the existentialist philosopher Jean-Paul Sartre. For him, freedom is not something we have but something we are. Sartre defends this position most clearly in *Existentialism Is a Humanism*, trans. Carol Macomber (New Haven: Yale University Press, 1970).
4. Matthew Stanley, *Einstein's War: How Relativity Triumphed amid the Vicious*

Nationalism of World War II (New York: Penguin Random House, 2019), 215–216.
5. Stanley, *Einstein's War*, 215–216.
6. Just to be clear, I am not denying that the laws of nature can also change or undergo an evolution over enormous periods of time, just as grammatical rules can change dramatically over long periods of human history. But this possibility does not affect the argument of this chapter. The analogy of laws to grammar, like all analogies, is imperfect.
7. Patricia Churchland, *Conscience: The Origins of Moral Intuition* (New York: Norton, 2019).
8. According to the archaeonomic linguist Stephen Pinker, the whole notion of human dignity is a "stupid idea." Pinker, "The Stupidity of Dignity," *New Republic*, May 28, 2008, https://newrepublic.com/article/64674/the-stupidity-dignity.
9. Stephen Cave, "There's No Such Thing as Free Will: But We're Better Off Believing in It Anyway," *The Atlantic*, June 2016, https://www.theatlantic.com/magazine/archive/2016/06/theres-no-such-thing-as-free-will/480750/.
10. Ibid.
11. A good example of a post-Einsteinian understanding of freedom as independent of nature can be found in the existentialist theology of Rudolf Bultmann (1884–1976), one of the most famous Christian theologians of the twentieth century. See Rudolf Bultmann, "The New Testament and Mythology," in *Kerygma and Myth by Rudolf Bultmann and Five Critics*, ed. Hans Werner Bartsch, trans. Reginald Fuller (New York: Harper Torchbooks, 1961), 1–44. Existentialist philosophy in the twentieth century also usually assumed the independence of freedom from nature.
12. Quoted in *The Ultimate Quotable Einstein*, ed. Alice Calaprice (Princeton, N.J.: Princeton University Press 2011), 336.

Chapter 10. Faith

Epigraph 1: Albert Einstein, *Ideas and Opinions*, trans. Sonja Bargmann (New York: Crown, 1994), 44–47.
Epigraph 2: Einstein, *Ideas and Opinions*, 26.
Epigraph 3: T. S. Eliot, "East Coker," III.
1. *Pistis*, according to the biblical scholar N. T. Wright, is the Greek term Saint Paul uses in referring to the Christian's "loyalty" to Christ. Wright, *Paul: A Biography* (San Francisco: HarperOne), 90.

2. Einstein, *Ideas and Opinions*, 38–40.
3. Albert Einstein, "Science and God," *Forum and Century* 83 (1930): 373.
4. Recent archaeonomic debunkers of religion, especially of Christianity, include the following: Richard Dawkins, *The God Delusion* (New York: Houghton Mifflin, 2006); Sam Harris, *The End of Faith: Religion, Terror, and the Future of Reason* (New York: Norton, 2004); Harris, *Letter to a Christian Nation* (New York: Knopf, 2007); Christopher Hitchens, *God Is Not Great: How Religion Poisons Everything* (New York: Hachette, 2007); Victor J. Stenger, *God, The Failed Hypothesis: How Science Shows That God Does Not Exist* (Amherst, N.Y.: Prometheus, 2007); Carl Sagan, *The Demon-Haunted World: Science as a Candle in the Dark* (New York: Ballantine, 1997); Steven Weinberg, *Dreams of a Final Theory* (New York: Pantheon, 1992); Michael Shermer, *How We Believe: The Search for God in an Age of Science* (New York: Freeman, 2000); Owen Flanagan, *The Problem of the Soul: Two Visions of Mind and How to Reconcile Them* (New York: Basic, 2002); Daniel Dennett, *Breaking the Spell: Religion as a Natural Phenomenon* (New York: Viking, 2006).
5. The clearest expression of archaeonomic faith that I have seen is Edward O. Wilson's book, *Consilience: The Unity of Knowledge* (New York: Knopf, 1998).
6. As an example of a pure archaeonomic faith, see once again Alex Rosenberg, *The Atheist's Guide to Reality: Enjoying Life without Illusions* (New York: Norton, 2012), 20–21. Yet another archaeonomic materialist is the physicist Sean Carroll; see his book *The Big Picture: On the Origins of Life, Meaning, and the Universe Itself* (New York: Dutton, 2016). See also David Papineau, *Philosophical Naturalism* (Cambridge, Mass.: Blackwell, 1993), 3.
7. Einstein, *Ideas and Opinions*, 262.
8. Quoted in Helen Dukas and Banesh Hoffman, *Albert Einstein, the Human Side: Glimpses from His Archives* (Princeton, N.J.: Princeton University Press), 66.
9. David E. Rowe and Robert Shulmann, *Einstein on Politics: His Private Thoughts and Public Stands on Nationalism, Zionism, War, Peace, and the Bomb* (Princeton, N.J.: Princeton University Press, 2007), 17. An anecdote illustrates the point: "At home in Berlin in April 1929, Albert Einstein received an urgent telegram from Rabbi Herbert S. Goldstein of New York: 'Do you believe in God? Stop. Answer paid 50 words.' Boston Archbishop William Henry Cardinal O'Connell had derided Einstein's famous relativity theories as 'befogged speculation' conjuring 'the ghastly apparition of Atheism.' An alarmed Goldstein sought to douse these rhetorical flames with reassurance from the great man himself. 'I believe in Spinoza's God,' Einstein wired back,

'Who reveals Himself in the lawful harmony of the world, not in a God Who concerns Himself with the fate and the doings of mankind.' The rabbi might have saved himself a little money; in the end, Einstein's reply in the original German used only 25 words." Mandy Katz, "Einstein and His God," *Moment*, April–May 2007, https://momentmag.com/einstein-and-his-god/.
10. This misfortune is what Teilhard meant when he said that Christianity henceforth needs to be liberated from its Mediterranean environment. See Pierre Teilhard de Chardin, *The Divine Milieu* (New York: Harper and Row, 1965), 46.
11. See Pierre Teilhard de Chardin, *Human Energy*, trans. J. M. Cohen (New York: Harcourt Brace Jovanovich, 1962), 29.
12. From Einstein's speech "Geometry and Experience" to the Prussian Academy of Sciences in 1921, published in Albert Einstein, *Sidelights on Relativity* (Garden City, N.Y.: Dover, 2010), 14.
13. No less interesting than the new cosmic story is the story of how science itself has awakened to the fact that the universe is a story. The history of how scientists and philosophers like Einstein, sometimes in spite of their private prejudices, gradually discovered that the cosmos is a story is itself an interesting part of the larger drama of cosmic awakening. A readable depiction of both stories can be found in Timothy Ferris, *Coming of Age in the Milky Way* (New York: Harper Perennial, 1988).

Chapter 11. Hope

Epigraph 1: James Jeans, *The Mysterious Universe*, rev. ed. (1930; New York: Macmillan, 1948), 15–16.
Epigraph 2: Pierre Teilhard de Chardin, *Writings in Time of War*, trans. René Hague (New York: Harper and Row, 1968), 55–56.
1. Lee Smolin, *Time Reborn: From the Crisis in Physics to the Future of the Universe* (New York: Houghton Mifflin Harcourt, 2013), 4–11. Smolin seems to be somewhat of an exception among physicists in affirming the objective reality of time. By contrast, the popular physicist Carlo Rovelli may be more representative. He, like Einstein, concludes that our sense that time passes is nothing but an "effect of perspective." "The study of time," Rovelli says, "does nothing but return us to ourselves." Carlo Rovelli, *The Order of Time* (New York: Riverhead, 2018), 169.
2. General relativity, as summarized by the physicist John Wheeler, implies that "matter tells space how to curve, and space tells matter how to move." See K. C. Cole, *A Hole in the Universe: How Scientists Peered over the Edge of*

Emptiness and Found Everything (New York: Harvest, 2001), 35. Many scientists, not just Einstein, consider the timeless truths of geometry more real and effective than anything that happens in time.
3. Romans 8:20-23. See also Teilhard, *Writings in Time of War*, 58-59.
4. John 1:14; Colossians 2:9.
5. Colossians 1:17.
6. Acts (throughout).
7. Matthew 28; Mark 16; Luke 24; John 20-21.
8. Nicene Creed, based on Psalm 110:1.
9. Psalm 110:1; Matthew 28; Nicene Creed.
10. See Luke Johnson, *The Real Jesus: The Misguided Quest for the Historical Jesus and the Truth of the Traditional Gospel* (New York: HarperOne, 1997).
11. Since I am taking a cosmological and anticipatory theological perspective on Christian beliefs in this book, I cannot deal in depth here with questions regarding the historical facticity of specific features of the Easter story.
12. Bertrand Russell, *A Free Man's Worship* (Portland, Ore.: Mosher, 1923), 7.
13. Freeman Dyson, *Disturbing the Universe* (New York: Basic, 1981).
14. William James, *Pragmatism* (Cleveland, Ohio: Meridian, 1964), 76. James himself did not embrace the scientific materialism he is talking about here.
15. In this interpretation I have been following Jürgen Moltmann, *Theology of Hope: On the Ground and the Implications of a Christian Eschatology*, trans. James W. Leitch (New York: Harper and Row, 1967), 95-138.
16. Luke 12:7; Psalm 56:8.
17. Paul Tillich, *The Eternal Now* (New York: Scribner, 1963), 35.
18. Alfred North Whitehead, *Modes of Thought* (New York: Free Press, 1968), 127-169.
19. Whitehead, *Modes of Thought*, 127-169; For Whitehead's discussion of perishing see especially Alfred North Whitehead, *Process and Reality*, corrected ed., ed. David Ray Griffin and Donald W. Sherburne (New York: Free Press, 1968), 29, 81-2, 146-47, 347-51; and Alfred North Whitehead, "Immortality," in *The Philosophy of Alfred North Whitehead*, ed. Paul A. Schillp (Evanston, Ill.: Northwestern University Press, 1941), 682-700. See also Charles Hartshorne, *The Logic of Perfection* (Lasalle, Ill.: Open Court, 1962), 24-62, 250, 340-341, 346-351.
20. This is part of the summation of general relativity, mentioned elsewhere, by the physicist John Archibald Wheeler: "Matter tells space-time how to bend, and space-time tells matter how to move." See Marcelo Gleiser, *The Dancing Universe: From Creation Myths to the Big Bang* (Hanover, N.H.: Dartmouth College Press), 257.

21. Whitehead, *Process and Reality,* 346.
22. See Alfred North Whitehead, *Science and the Modern World* (New York: Free Press), 192.
23. Whitehead refers to this receptivity as the "Consequent Nature of God." Whitehead, *Process and Reality,* 343–351. In my references here to Whitehead I am not adhering exactly to his way of understanding nature and God, since I want to emphasize, more than he does, the cosmic process as dramatic. There is in Whitehead's thought a Platonic quality that lines up at times more closely with analogy than with anticipation.
24. Whitehead, *Process and Reality,* 67, 88, 164, and passim.
25. Alfred North Whitehead, *Adventures of Ideas* (New York: Free Press, 1967), 252–272, 283–295.
26. Pope Francis, Encyclical Letter "Laudato si' of the Holy Father Francis on Care for Our Common Home," #243, May 24, 2015 (emphasis added), http://w2.vatican.va/content/francesco/en/encyclicals/documents/papa-francesco_20150524_enciclica-laudato-si.html.
27. See "The Pastoral Constitution on the Church in the Modern World," in *The Documents of Vatican II,* ed. Walter M. Abbott (New York: Guild Press, 1966), 203–204, 218, 233.
28. This is not a denial of divine "immutability." What remain immutable, eternal, and absolute in God are the infinite compassion and never exhausted fidelity that move God to become incarnate in matter and time.
29. Furthermore, there is no good reason, theologically, to assume that the "subjectivity" of other living beings is not also saved and transformed within the everlasting compassion of God.

Chapter 12. Compassion

Epigraph 1: Albert Einstein, *New York Times,* November 9, 1930, quoted in Nancy Frankenbury, ed., *The Faith of Scientists in Their Own Words* (Princeton, N.J.: Princeton University Press), 56.

Epigraph 2: Holmes Rolston III (interview with Jeff Dodge), "Q&A with Holmes Rolston: Life Persists in the Midst of Its Perpetual Perishing," *Source* (Colorado State University), May, 2020, https://libarts.source.colostate.edu/qa-with-holmes-rolston-life-persists-in-the-midst-of-its-perpetual-perishing/.

1. Pierre Teilhard de Chardin, *Christianity and Evolution,* trans. René Hague (New York: Harvest Books, 1974), 79–95, 84–86, 131–132.
2. See George Williams, "Mother Nature Is a Wicked Old Witch!" in *Evolutionary*

Ethics, ed. Matthew H. Nitecki and Doris Nitecki (Albany: State University of New York Press, 1995), 217–231.
3. Philip Kitcher, *Living with Darwin: Evolution, Design, and the Future of Faith* (New York: Oxford University Press, 2009), 124.
4. Stephen J. Gould, "Introduction," in Carl Zimmer, *Evolution: The Triumph of an Idea from Darwin to DNA* (London: Arrow, 2003), xvi–xvii. Here I am building on thoughts I expressed earlier in a chapter titled "Wrongness" in *The New Cosmic Story: Inside Our Awakening Universe* (New Haven: Yale University Press, 2017), 159–173.
5. Pierre Teilhard de Chardin, *Activation of Energy,* trans. René Hague (New York: Harcourt Brace Jovanovich, 1970), 231–243. The interpretation of evil in this chapter owes much to the works of Teilhard cited above, especially *Christianity and Evolution.*

Chapter 13. Caring for Nature

Epigraph: Pope Francis, Encyclical Letter "Laudato si' of the Holy Father Francis on Care for Our Common Home," #243, May 24, 2015, http://w2.vatican.va/content/francesco/en/encyclicals/documents/papa-francesco_20150524_enciclica-laudato-si.html.

1. For more extended treatments of this question see my books *The Promise of Nature* (New York: Paulist, 1993) and the more recent *Resting on the Future: Catholic Theology for an Unfinished Universe* (New York: Bloomsbury, 2015), 149–158. This chapter condenses, revises, and adapts to the aims of the present book some material from my recent essay "The Unfinished Sacrament of Creation: Christian Faith and the Promise of Nature," in *Ecotheology, a Christian Conversation,* ed. Kiara A. Jordan and Alan G. Padgett (Grand Rapids, Mich.: Eerdmans, 2020), 165–189 (used with permission).
2. An early discussion of the recent ecological awakening among Protestant theologians is the fine book by Paul Santmire, *The Travail of Nature* (Philadelphia: Fortress, 1985); among Catholics an early presentation is that of the priest and pastor Charles M. Murphy, *At Home on Earth* (New York: Crossroad, 1989).
3. For example, John Passmore, *Man's Responsibility for Nature* (New York: Scribner, 1974), 184.
4. See Thomas Berry, *The Dream of the Earth* (San Francisco: Sierra Club Books, 1988); also Charlene Spretnak, *States of Grace* (San Francisco: Harper and Row, 1991). Lynn White Jr. in his essay "The Historical Roots of Our Ecological Crisis," *Science* 155, no. 3767 (1967): 1203–1207, indicted Christianity for interpreting the biblical theme of dominion as justification for eco-

nomic progress that imperils the natural world. As an alternative, he prefers Benedictine sacramental (analogical) theology to the anticipatory theme of biblical hope.
5. See especially Jürgen Moltmann, *Theology of Hope,* trans. James Leitch (New York: Harper and Row, 1967; and Jürgen Moltmann, *The Experiment Hope,* ed. and trans., M. Douglas Meeks (Philadelphia: Fortress, 1975). Moltmann's theology of hope is deeply influenced by the Marxist philosopher Ernst Bloch's three-volume work, *The Principle of Hope,* trans. Neville Plaice, Stephen Plaice, and Paul Knight (Oxford: Basil Blackwell, 1986).
6. See Jürgen Moltmann, *God in Creation,* trans. Margaret Kohl (San Francisco: Harper and Row, 1985).
7. N. T. Wright, *Surprised by Hope* (San Francisco: HarperOne, 2008).
8. That the universe has the capacity to bring forth "fuller being" or "more being" in the future is a major theme in the writings of Pierre Teilhard de Chardin, as listed earlier. In this sense, Teilhard's religious and cosmic vision is indispensable to a Christian ecological theology.
9. As noted earlier, a good example of archaeonomic materialism may be found in the writings of E. O. Wilson of Harvard University. See especially his book *Consilience: The Unity of Knowledge* (New York: Knopf, 1998). For Wilson, the principle of coherence that holds everything in the universe together lies not in the present or future but in the remote cosmic past before life existed. Along with other archaeonomic thinkers, Wilson holds intellectually that the integrating discipline in our universities is not ecology but physics.
10. Edward O. Wilson, *Biophilia* (Cambridge: Harvard University Press, 1984).
11. Passmore, *Man's Responsibility for Nature,* 184.
12. See Carl Sagan, *Cosmos* (New York: Ballantine, 1985), 1. Also see Charley Hardwick, *Events of Grace: Naturalism, Existentialism, and Theology* (Cambridge: Cambridge University Press, 1996).
13. A good example of the sacramental approach to ecological responsibility is offered in an article by Michael J. and Kenneth R. Himes, "The Sacrament of Creation," *Commonweal* 117 (January 1990): 42–49.
14. Jürgen Moltmann and Teilhard de Chardin have both written important interpretations of that view.
15. This is a point made often in the writings of Pierre Teilhard de Chardin—for example, in *Human Energy,* trans. J. M. Cohen (New York: Harvest Books/Harcourt Brace Jovanovich, 1962), 29; and *Activation of Energy,* trans. René Hague (New York: Harcourt Brace Jovanovich, 1970), 229–244.
16. Romans 8:22.

17. Identifying hope as a fundamental ecological virtue is a point implicit in Pope Francis's encyclical letter "Laudato si'."
18. Teilhard refers to God as the future on which the world rests as its sole support in *Activation of Energy*, 139, 239. I have developed this theme at book length in *Resting on the Future: Catholic Theology for an Unfinished Universe* (New York: Bloomsbury, 2015).

Conclusion

1. See Paul Tillich, *The Courage to Be* (New Haven: Yale University Press, 1953).
2. 1 Corinthians, 15:27–28.
3. Romans 8:24.
4. Paraphrasing Jürgen Moltmann, *Theology of Hope: On the Ground and the Implications of a Christian Eschatology*, trans. James W. Leitch (New York: Harper and Row, 1967), 26–32.
5. Moltmann, *Theology of Hope*, 16.

Index

Abraham (biblical figure), 69, 75, 79, 91, 126, 155, 162, 194. *See also* anticipation; patience
Absolute Future, 51
acceleration, 39–40, 72
achievement, logic of, 99–100
actual, vs. ideal, 173
adaptation, 117, 137. *See also* evolution
adventure, 167
analogy, 69, 71–72, 74, 75, 76, 77, 78, 79, 91; appeal of, 88; archaeonomists' use of, 119; in Christian spirituality, 126; compassion and, 183; determinism and, 139; divine care and, 165; ecological responsibility and, 184, 186, 190–93; Einstein's, 72, 119, 126, 127, 154; ethical life and, 188–89; evil and, 180, 181–82; faith and, 148–49; freedom and, 135, 136, 139–40, 142; hope and, 161–62, 169–70, 193; impatience in, 201; intelligibility in, 151; life and, 102–7, 108; meaning in nature and, 187; mind/thought and, 118, 121, 122, 124; morality and, 182–83; Nicene Creed and, 162, 169; optimism of, 155, 156, 161, 163; origins and, 87–90; rejection of, 124, 127; resurrection and, 159, 162; science and, 72; suffering and, 179; theology and, 76, 161; time and, 89, 142, 152, 156, 161, 181–82, 194; timelessness in, 186, 191–92; in Western theology, 76. *See also* myths
anticipation, 66, 69, 72–75, 77, 78–79, 91, 102; after Einstein, 121, 179; Big Bang cosmology and, 122; compassion and, 175, 180; divine care and, 165; ecological responsibility and, 185–87, 192–
94, 195; evil and, 180, 183; evolution and, 122; faith and, 75, 78, 149–50, 151, 154; freedom and, 135–36, 140–42, 143; hope and, 75, 162–63, 169, 200–201; intelligibility and, 73, 74, 92, 153; life and, 97, 99, 107–11, 112; meaning of, 94; mind/thought and, 120–24, 125, 126, 128; multiverse and, 110–11; nature's cumulative dramatic value in, 194; Nicene Creed and, 78, 126–27; origins and, 90–92; past and, 73–74; patience in, 201; relativity's compatibility with, 122; resurrection-hope and, 159; science and, 111, 122, 163. *See also* patience; waiting
antigravity factor, 82
antitheism, Einstein's, 151–52
Apostles' Creed, 30
Aquinas, Thomas, 105
archaeonomy, 68–71, 74, 75–76, 77, 78, 79, 90–91; after Einstein, 123; analogy used in, 119; anxiety about perishing and, 85; compassion and, 178–80, 182; continuity of nature with physical past in, 124; cosmic immensities and, 107; Dennett, 115–16, 117, 119; ecological responsibility and, 187–90; Einstein's, 126, 127, 154; evolution and, 116–17; faith and, 147–48, 152; Flanagan, 116, 117, 119; freedom and, 136–39, 141–42; hope and, 159–61, 168; impatience of, 160–61, 163, 201; intelligibility and, 151, 153; life and, 96–97, 98, 100–102, 103–4, 106, 109; vs. materialism, 71; mind/thought and, 114–18, 119, 122, 126, 127, 147; moral consciousness and, 176–77; origins and, 85–87;

archaeonomy (*continued*)
  pessimism of, 155, 156, 169; prevalence of, 127; rejecting, 124, 127; resurrection narratives and, 159; time and, 142, 160–61, 181, 182, 194; waiting and, 147. *See also* materialism; pessimism
atheism, 3, 26
atomism, 69, 76, 90, 136–37, 165. *See also* archaeonomy
Augustine, Saint, 191
awakening, 4, 29, 36, 43–44, 154. *See also* not-yet; universe, unfinished

background, awareness of, 1
balance, 123
beauty, 63, 167–68. *See also* values
beginning, 28, 81, 85. *See also* creation; origins
being. *See* life
Bergson, Henri, 18, 106–7, 109
biblical literalists, 45–46
Big Bang, 8, 36, 54, 80, 93, 108; anticipation's compatibility with, 122; beginning and, 28, 81; gravity and, 32; life and, 110; relativity and, 3–4, 19, 24, 32, 41; time and, 142. *See also* relativity; universe, unfinished
biology, 100–101
Bonaventure, Saint, 61
brahman, 10
Buddhism, 10, 178
burial practices, 172–73

care, divine, 165
Carroll, Sean, 87–88
Catholicism, 105, 107. *See also* Christianity
Cave, Stephen, 137–38
children, 57
Christ, imitation of, 200. *See also* Jesus
Christianity: after Einstein, 179; analogy and, 126; archaeonomists on, 147; belief in God, 11; Catholicism, 105, 107; ecological responsibility and, 185, 186, 189; faith in, 144; future and, 76; mystery in, 51; sensitivity to suffering in, 178. *See also* faith; theology
Churchland, Patricia, 137
Clough, Arthur Hugh, 157
cognition, 98, 118–19. *See also* mind/thought
coherence, 40, 41; vs. geometry, 168; longing for, 150; resurrection and, 168. *See also* intelligibility; unity
Colossians, 157
compassion, 171; analogy and, 183; anticipation and, 175, 180; archaeonomy and, 178–80, 182; as cosmic development, 171–72, 175–78; ecological importance of, 185; emergence of, 172–73; evolution of, 174–78; God's, 57, 164, 168, 170; impatience and, 178–80. *See also* ethics
complexity, 36, 95–96, 107, 171, 172. *See also* immensities, cosmic; life; mind/thought
comprehensibility. *See* coherence; intelligibility; unity
consciousness, 36, 93, 114. *See also* mind/thought
Consciousness Explained (Dennett), 115–16
conservation, environmental, 195. *See also* ecological responsibility
consistency, 47, 66
consubstantiation, 13
contingency, 26
cosmogenesis. *See* beginning; Big Bang; origins
cosmos, 47; cosmological constant, 82; cosmological perspective, 9, 185. *See also* universe
Council of Chalcedon, 196
Council of Nicea. *See* Nicene Creed
creation, 4; Big Bang and, 80; creator God, 80, 81, 82; divine love and, 92;

Index

freedom and, 130, 134 (*see also* freedom: infinite); in Genesis, 34; kenosis and, 130; in Nicene Creed, 43; plausibility of, 83; as unfinished drama, 28 (*see also* universe, unfinished)
creator, uncreated, 80
Crick, Francis, 100–101
crucifixion, 14, 30, 196–97
culture, entropy and, 84

dao, 10
dark energy, 82
darkness, 83
Darwin, Charles, 3; age of life and, 34; analogy after, 73, 105; Christian faith and, 26, 27; dramatic coherence and, 40, 66, 67; suffering and, 174, 175, 176. *See also* evolution
de Sitter, Willem, 113
death, 9; in fixed universe, 1; inability to make sense of, 179. *See also* perishing
deception, 117
Democritus, 69, 75, 79, 122, 126, 156, 165. *See also* archaeonomy
Dennett, Daniel, 60, 115–16, 117, 119
Descartes, René, 59, 119, 152
despair, 173
determinism, 23–24, 38, 60, 135; analogy and, 139; concealed from masses, 139; Einstein's, 123, 136–37; focus on geometry of relativity, 132–34, 143; freedom and, 131–34, 136; God and, 137; as illusion, 136; morality and, 182; mystery and, 62; science and, 137–39; Spinoza and, 136; universe's, 60. *See also* laws/lawfulness
dignity, human, 130, 131–32, 137
dinosaurs, 42
disaster, 27
Docetism, 29, 30–31
Dostoyevsky, Fyodor, 139

drama: Einstein's failure to appreciate, 123; universe as, 59. *See also* awakening
Dreams of a Final Theory (Weinberg), 123
dualism, 119, 125; freedom and, 136; mind/thought and, 122–23; rejecting, 127
Dyson, Freeman, 159

Easter event. *See* resurrection
ecological responsibility, 167–68; analogy and, 184, 186, 190–93; anticipation and, 185–87, 192–94, 195; archaeonomy and, 187–90; Christianity and, 185, 186, 189; cosmological perspective, 185; gratitude and, 192; indifference to, 157; materialism and, 188; Nicene Creed and, 186; pragmatic reasons for, 189–90; stewardship, 195. *See also* nature
education, religious, 55
Edwards, Jonathan, 71–72
Einstein, Albert: his importance for theology, 4, 15, 199–200; reflections on, 3; religious sensibilities of, 25; use of analogical thinking, 119. *See also* gravity; relativity
Eliade, Mircea, 84
Eliot, T. S., 144
energy: dark, 82; loss of, 28, 36, 44, 83–84, 86, 156
entropy, 28, 36, 44, 83–84, 86, 156
environment, 184, 195. *See also* ecological responsibility
Epicurus, 10
equivalence, 56–57, 58, 59, 60, 123
eternity, 9; Christians' love of, 29–31; cleansing of time, 30; Einstein's love of, 16, 18, 20, 32, 50, 65, 72, 87, 141, 199; Einstein's understanding of, 12; in Nicene Creed, 18; pantheism and, 21–23; time's relationship with, 13, 104. *See also* everlastingness, divine; time; timelessness

ethics, 172, 188–89. *See also* compassion; morality
Ethics, The (Spinoza), 59
everlastingness, divine, 11, 17. *See also* eternity; time; timelessness
evil, 9, 93, 141, 173–74; analogy and, 181–82; anticipation and, 183; inability to make sense of, 179; origins of, 173; reality of time and, 179, 180–82; reconciling God's existence with, 180–82; in stories, 41
evolution, 2, 26, 27, 48, 111, 172; anticipation's compatibility with, 122; mind/thought and, 116–18; as movement toward unity, 109; religion's compatibility with, 3; suffering in, 174–75, 176. *See also* Darwin, Charles; humans; life; natural selection
excess, divine, 55–61
expectation, 74–75, 150. *See also* anticipation; patience; waiting
experiencing, 114. *See also* mind/thought
exploration, mystery and, 50

failure, 98, 99, 100. *See also* perishing
faith, 144; after Einstein, 179; allowing for, 153; analogy and, 148–49, 156; anticipation and, 75, 78, 149–50, 151, 154; archaeonomy and, 147–48, 152; in Christian religion, 144; in classical theology, 149; Einstein and, 21, 144–45, 148, 149; forms of, 147; in Nicene Creed, 144; patience/waiting and, 47, 149, 151; profession of, 78 (*see also* Nicene Creed); science and, 21, 145–46; smallness of, 55; striving of life and, 146; time and, 91, 149–50, 151, 152; universe's meaning and, 154. *See also* religion; theology
fatalism, 169
finalism, 74

First Council of Constantinople, 196
Flanagan, Owen, 116, 117, 119
forgiveness, 57, 134
fortune, 27
fossil record, 42, 172
fourth dimension, 36. *See also* time
Francis (pope), 167–68, 184, 185
freedom, 23; analogy and, 135, 136, 139–40, 142; anticipation and, 135–36, 140–42, 143; archaeonomy and, 136–39, 141–42; of choice, 130–31, 137–38; concealed from masses, 138, 139; creation and, 130, 134; denial of, 129–30, 132, 137–39, 141; determinism and, 131–34, 136; emergence of, 141; fixed universe and, 181; geometry and, 143; human dignity and, 130, 131–32; as illusion, 129–30, 137–38; infinite, 130, 132, 135; lawfulness and, 129–31, 132, 135, 143; meanings of, 130; relativity and, 130–31; science and, 132–34, 135; soul's need for, 76; theology and, 134–35; time and, 142
Friedman, Alexander, 113
fulfillment, promise of, 163, 199. *See also* hope
future, 11, 12, 20, 78, 90, 92, 107, 161; Christianity and, 76; existence of life and, 109; fixed universe and, 181; need for, 76; power of, 94. *See also* anticipation; not-yet; waiting

galaxies, number of, 54–55
Galileo Galilei, 1, 152
Gamow, George, 3
Genesis, 34, 80, 81, 134
geometry, 37, 133, 149; vs. dramatic coherence, 168; focus on, 58, 91–92, 123, 132–34, 143, 165; freedom and, 143; intelligibility and, 27, 28, 29, 40, 41, 47, 50; relativity and, 154; Spinoza's idealization of, 59; timelessness of,

Index

166; time's subordination to, 73, 152; unfinished universe and, 65; unifying power of, 72

Gnosticism, 29

God, 9, 10, 166; belief in, 11; compassion of, 57, 164, 168, 170; as creator, 80, 81, 82 (*see also* creation); denial of reality of, 142; determinism and, 137; Einstein on, 20–21; of excess, 56, 57; in fixed universe, 1; identified with nature/universe, 21, 148 (*see also* pantheism); as illusion, 137; incarnation of, 170; matter and, 30, 159, 170; mystery of, 50, 51, 61, 91; new, 8; as not-yet, 11, 12, 14, 17; oneness of, 151; past and, 164–69, 170; personal, 6, 51, 64–65, 78, 129, 132, 152, 166; promise of, 11, 162; science and, 25–26; Spinoza's, 18, 21, 24, 199; time and, 11–12, 17, 30, 159, 170; timelessness of, 25; transcendence of, 47. *See also* faith

God after Darwin (Haught), 3, 27

Good, the, 10

goodness, 63. *See also* values

Gospels: Gospel of John, 127, 157; inconclusive endings of, 159

Gould, Stephen Jay, 177

grammar, 133–34. *See also* geometry

gratitude, 192

gravity, 4, 19, 31, 32, 35, 36, 40, 72, 81

healing, 180, 187

heaven, ascension into, 169–70

heresies, 29–31

Holy Spirit, 103, 111

hope, 10, 11, 17, 41, 89, 150; analogy and, 161–62, 169–70, 193; anticipation and, 75, 162–63, 169, 200–201; archaeonomy and, 159–61, 168; awakening and, 4–5; divine memory and, 164–65; ecological importance of, 185; entrance of God into time and, 11–12;

misrepresentation of, 170; in Nicene Creed, 155–56, 157, 169–70, 173, 180, 197–98; promise of final fulfillment, 163; reasonableness of, 16; reasons for, 4, 5, 169; spread of, 158. *See also* resurrection

Hubble, Edwin, 3, 113

humans, 42, 54, 93. *See also* evolution; life; mind/thought

humility, 150, 185

ideal, 10, 173

immensities, cosmic, 95, 96, 97, 99, 101–2, 107, 110–11. *See also* complexity; numbers, immensity of; space; time

immensity: sense of, 53–55; spatiotemporal, 56

impatience, 23–27, 147, 153, 163, 201; of archaeonomy, 160–61, 163, 201; compassion and, 178–80; demand for magic and, 46. *See also* patience

infinity, 54, 192

intelligent design, 26, 27, 45

intelligibility, 21, 28; in analogy, 151; anticipation and, 73, 74, 92, 153; archaeonomy and, 120, 151, 153; dramatic, 27, 58–59 (*see also* universe, unfinished); geometry and, 47; life and, 97, 112; meaning distinguished from, 124; of multiverse, 111–12; mystery and, 50, 51–52; of nature, 38, 40; in not-yet, 44; seeking, 28–29; as simplification, 61; stories and, 41; time and, 19, 27; trust in one's own mind, 119. *See also* coherence; unity

Irenaeus of Lyons, 15

Israel. *See* Judaism

James, William, 160

Jeans, James, 155

Jesus, 91; death of, 14, 30, 196–97; historical, 30; rightness and, 10; as Son of God, 15, 29–31, 196; time and, 13, 29–31. *See also* resurrection
Jones, Peter, 70
Judaism: indestructible rightness in, 10–11; renewal of cosmos in, 170
justice, 10

Kant, Immanuel, 68, 153, 156, 172
kenosis, 14, 130. *See also* love: self-emptying
Kitcher, Philip, 175, 179

Laozi, 10
laws/lawfulness, 36–38, 59, 60, 126; Einstein's focus on, 123; freedom and, 129–31, 132, 135, 143; idea of, 58; interventions in, 64–65, 152, 168; life and, 125; meaning and, 134; mind/thought and, 125; morality and, 137; of nature, 36–38. *See also* determinism
Lemaître, Georges, 3, 36, 48, 82, 113
life, 99; age of, 34, 42; analogy and, 102–7, 108; anticipation and, 97, 99, 107–11, 112; archaeonomy and, 96–97, 98, 100–102, 103–4, 106, 109; Big Bang and, 110; distinguished from nonlife, 100; divergence of, 109; as drama, 99; emergence of, 146, 169; entropy and, 84; fixed universe and, 181; intelligibility and, 97, 112; lawfulness and, 125; matter's discontinuity with, 103; in Nicene Creed, 95, 103, 111; origins of, 42–43, 111; point of, 88–89; sacramental appreciation of, 105; striving of, 97–99, 100, 101, 141, 146, 201; success/failure in, 98, 99; as unfinished story, 26; vitalism, 105–7, 108, 112, 120, 125, 189. *See also* complexity; creation; evolution; mind/thought; suffering
lifelessness, 100, 101

light: speed of, 39; from stars, 82–83
logic: of achievement, 99–100; of equivalence, 58, 59
logos, 10
Lord's Supper, 159
love, 14; self-emptying, 130; self-giving, 92; selfless, 13

Mack, Katie, 37
Macquarrie, John, 13
magic, 43, 45, 46
materialism, 75, 79; after Einstein, 123; vs. archaeonomic approach, 71; ecological responsibility and, 188; escaping time in, 35, 61; ethical life and, 188; logic of equivalence, 60; mind/thought and, 114–18, 119–20, 121; mystery and, 62; narrative quality of nature ignored by, 45–46, 108; reduction of life to nonlife by, 23–24, 54, 101, 102–3; rejection of, 124, 127. *See also* archaeonomy
matter: God and, 30, 159, 170; inherent narrativity of, 28; life's discontinuity with, 103; time's relationship with, 16, 17, 19, 28, 36, 37, 39, 199–200
meaning, 9, 27, 41, 43–44, 65–79, 114, 200; craving for, 40; fixed universe and, 1, 181; grammar and, 133–34; intelligibility distinguished from, 124; lack of, 114, 123; laws and, 134; patience and, 66; searches for, 46, 48, 55; stories and, 28; in time's passage, 67. *See also* intelligibility
mechanism, 133
memory, 163–69
metaphysics, conflated with physics, 88
method, scientific, 29
Milky Way, 54
mind/thought, 8–9, 113; analogy and, 118, 121, 122, 124; anticipation and, 120–24, 125, 126, 128; archaeonomy and, 114–18, 119, 122, 126, 127, 147; awaken-

ing and, 125–26; emergence of, 70, 95, 114, 146, 171, 172; entropy and, 84; evolution and, 116–18; existence of, 38–39, 101; hard problem of, 126; as intensification of drama of awakening, 127; lawfulness and, 125; materialism and, 114–18, 119–20, 121; myths and, 118–19; nature disconnected from, 119, 121, 123, 125–26; origin of, 42–43, 122; as part of nature/universe, 38–39, 124; reduced to mindlessness, 114, 120, 127, 147; trust of, 115–17, 118, 119, 121–22, 124, 127–28, 147–49. *See also* compassion; complexity; life; morality

Minkowski, Herman, 34
Moltmann, Jürgen, 14
Monod, Jacques, 96–97, 102, 109, 110
Monophysitism, 29
monotheistic theology, 151–52
moral consciousness, 176–77
morality, 138, 141, 148; analogy and, 182–83; contributing to transformation of universe and, 154; determinism and, 182; lawfulness and, 137. *See also* ethics
Moses (biblical figure), 91, 162
multiverse, 55, 88, 93–94, 110–11, 112, 127
mystery, 20; as Absolute Future, 51; awareness of, 49; children and, 57; in Christian faith, 51; determinism and, 62; Einstein and, 49–50, 149; of God, 50, 51, 61, 91; indestructibility of, 50, 51; inexhaustibility of, 52–53, 54; intelligibility and, 50, 51–52; as name for God, 61; as not-yet, 51; as personal, 53; vs. problem, 50; reality of, 49; religion as response to, 50; science and, 49, 61–62; of superabundance, 55–61; superpersonal attributes, 53; in theology, 54
mysticism, medieval, 53

myths, 41; cognition and, 118–19; coherence and, 150; evil in, 173; origins and, 84–85. *See also* analogy

Nagel, Thomas, 118
narrative cosmological principle, 28
narratives, 41. *See also* stories
natural selection, 26, 174–75, 176. *See also* evolution
nature, 16, 40; as awakening, 7, 28; caring for, 184; continuity with physical past, 124; cumulative dramatic value of, 194; as design vs. drama, 46, 48; entropy and, 28; God identified with, 21, 148 (*see also* pantheism); history of, 69 (*see also* archaeonomy); indifference to concern for, 157; intelligibility of, 38, 40; interventions in, 64–65, 78, 152, 168; laws of, 36–38; levels of understanding of, 153; love of, 33; mind as part of, 124; mind disconnected from, 119, 121, 123, 125–26; narrative quality of, 45, 79, 120 (*see also* intelligibility; time; universe, unfinished); as own cause, 24; perishability of, 191; as promise, 195; revelatory value of, 191; Spinoza and, 20; spirituality and, 8; time and, 26, 40; wrongness and, 175. *See also* creation; ecological responsibility; pantheism; universe; universe, unfinished
neuroscience, 137–39. *See also* science
New Testament, resurrection narratives in, 158–59, 162–63
Newton, Isaac, 1, 39
Nicene Creed, 6, 13; affirmation of oneness of God in, 151; analogical interpretation of, 162, 169; anticipatory reading of, 78, 126–27; assumption of indestructibility, 197–98; background of, 196; caring for nature and, 186; Catholic interpretations of,

Nicene Creed (*continued*)
105; church in, 198; contradiction to life's suffering in, 180; creation in, 43; disputes about, 196; eternity in, 18; expectation and, 150; faith in, 144; hope in, 155–56, 157, 169–70, 173, 180, 197–98 (*see also* resurrection); identity of Jesus as Son of God in, 196; inseparability of God, matter, and time in, 159, 170; life in, 95, 103, 111; origins in, 88; pantheism rejected in, 22; superabundance in, 58, 61; text of, 201–2; time in, 15, 30, 31, 32–33, 197–98

nighttime, 83

not-yet, 16, 24, 29, 41; God as, 11, 12, 14, 17; intelligibility in, 44; mystery as, 51; of resurrection, 158. *See also* awakening; future; universe, unfinished

novelty, 167

numbers, immensity of, 110–11. *See also* immensities, cosmic

Olbers' paradox, 82–83

optimism, 155, 156, 161, 163

origins, 80, 93; analogy and, 87–90; anticipation and, 90–92; archaeonomy and, 85–87; attraction to, 84; religion/myths and, 84–85; return to, 85; theology and, 90. *See also* beginning; creation

overflowing, images of, 55–61

Pagels, Heinz, 52

pantheism, 18, 21–23, 24, 25, 26, 29, 30, 87, 147, 148

past: anticipation and, 73–74; God and, 164–69, 170; metaphysics of (*see* archaeonomy); repository of, 163–69

patience, 67, 74, 89, 92, 122, 149, 150, 163; in anticipation, 201; atrophy of, 75–79; divine, 46; faith and, 47, 149, 151; in quest for meaning, 66. *See also* anticipation; impatience; waiting

Paul (apostle), 157, 158, 194, 200–201

perennialism, 103, 189

perfection, 48, 76, 180

perishing, 11, 15, 173; analogy's appeal and, 88; archaeonomy and, 85; cleansing eternity of time and, 30; comfort for anxiety about, 86–87; Einstein's sensitivity to, 87; of nature, 191; seeking return to origins and, 84–85; solution to, 77; of time, 166 (*see also* time, irreversible); timelessness's allure and, 156–57; of universe (*see* universe: collapse of)

pessimism, 147, 155, 156, 160, 169, 184. *See also* archaeonomy

philosophy, 126; James, 160; Kitcher, 175, 179; objective of, 116; Russell, 159; Whitehead, 34, 95, 164, 165–68, 170. *See also* analogy; archaeonomy; Democritus; Plato

physics, 88. *See also* analogy

Plato, 10, 69, 71, 75, 79, 89, 122, 125, 126. *See also* analogy

Platonic thought, 32, 46, 47, 89, 118, 156; mind/thought and, 121; rejecting, 124; on timelessness of God, 25. *See also* analogy

Polanyi, Michael, 97, 99

Pontius Pilate, 30

possibilities, 167

power, divine, 14

predictability, predictions, 23, 64. *See also* pantheism

problem, hard, 126

promise, 11, 93, 195. *See also* hope

purpose. *See* meaning

rationality, 117

realism, anticipation as, 78

reality, 50–51, 61–62

Index

redemption, 173
reductionism, 117. *See also* archaeonomy; materialism
regularity, lawful, 26. *See also* laws/lawfulness
relativity: anticipation's compatibility with, 122; Big Bang cosmology and, 3–4, 19, 24, 32, 41; conjoining of matter and time in, 28, 36, 37; cosmic immensities and, 108; freedom and, 130–31; geometry and, 154; mathematical interpretations of, 113; time and, 28, 36, 37, 39–40; unfinished universe and, 7, 38, 59, 154. *See also* Big Bang
religion: cosmological perspective on, 9; disillusionment with, 55; Einstein and, 64, 148; fresh thoughts about, 9; narratives in, 41; origins and, 84–85; past and, 164; as response to mystery, 50; science's compatibility with, 2, 64. *See also* Christianity; faith; rightness, indestructible; theology
remembering. *See* past
resurrection, 197, 198; analogy and, 159, 162; coherence and, 168; narratives, 158–59, 162–63; not-yet of, 158; resurrection-hope, 155–56, 157, 159, 162, 180; time and, 157–58; witnessing, 158. *See also* hope
Ricoeur, Paul, 58
rightness, indestructible, 9–15, 32, 84, 92; aspiring to, 36; in Buddhism, 10; divine superabundance and, 56; pantheism's ideal of, 23; protecting, 29; time and, 13, 15, 16, 17; Trinitarian theology and, 31–32. *See also* values
Rilke, Rainer Maria, 63
rituals, 172–73
Rolston, Holmes, III, 171
Rovelli, Carlo, 36
rules of nature. *See* laws/lawfulness
Russell, Bertrand, 159

Sagan, Carl, 190
salvation, personal, 2
science, 6, 21, 28–29, 54, 60; after Einstein, 9; analogy and, 72; anticipation and, 111, 122, 163; commitment to truth in, 145–46; determinism and, 137–39; faith and, 21, 145–46; freedom and, 132–34, 135; ideas about God and, 25–26; mystery and, 49, 61–62; neuroscience, 137–39; religion's compatibility with, 2, 64; revelatory task of, 65; theology's relationship with, 68; time and, 47, 157
Second Vatican Council, 170
self-creation, 24
self-identity, 135. *See also* freedom
self-worth/dignity, 130, 131–32, 137
simplification, mathematical, 60–61
sin, 173–74. *See also* evil
Smilansky, Saul, 138
Smolin, Lee, 16–17, 34, 95–96
space: immensity of, 56, 110–11; linked to drama of life's emergence, 107. *See also* immensities, cosmic
spacetime, true nature of, 131
species, 93
Spinoza, Baruch, 30, 41, 43, 164; determinism and, 136; on futility of waiting., 25; geometry idealized by, 59; God of, 18, 21, 24, 199; influence on Einstein, 52, 72, 92, 119, 132; love of eternity, 32; nature and, 20; pantheism of, 87, 148, 166; rejection of, 22, 77; time and, 39, 198
Spinoza Society of America, 129
spirituality, nature and, 8
Stanley, Matthew, 130–31, 134
stars, 27–28, 54, 82–83. *See also* universe
stewardship, 195
stories, 41; elements of, 66–67; function of, 40; intelligibility and, 41; meaning and, 28; sense of transcendent rightness and, 10; time and, 40

striving of life, 97–99, 100, 101, 141, 146, 201
subjectivity, 114. *See also* mind/thought
success, 98, 99–100
suffering, 9, 15, 93, 173, 177; in Christianity, 178; in fixed universe, 1; inability to make sense of, 179; in natural selection, 174–75, 176; Nicene Creed as contradiction to, 180
superabundance, mystery of, 55–61
superstition, 51
Swinburne, Algernon Charles, 86–87

Teilhard de Chardin, Pierre, 47, 80, 90, 92, 109–10, 155
teleology, 74
temperance, 150. *See also* virtue
theodicy problem, 180. *See also* evil
theology: after Einstein, 8, 9; analogy and, 76, 161 (*see also* analogy); defined, 7–8; dramatic coherence and, 40; Einstein's importance for, 4, 15, 199–200; end of universe and, 83; entropy and, 83–84; freedom and, 134–35; monotheistic, 151–52; mystery in, 54; narrative meaning and, 40; objectives of, 4, 65, 85; origins and, 90; perfection and, 48, 76; physics conflated with, 88; rightness identified in, 10; science and, 19, 68; time and, 15, 156. *See also* faith; religion
theology, classical: analogy and, 71–72; faith in, 149; infinite freedom in, 132
theology, medieval, 119, 125. *See also* analogy
"There's No Such Thing as Free Will" (Cave), 137–38
thermodynamics, 44. *See also* entropy
Thomism, 105
thought. *See* mind/thought
Tillich, Paul, 164–65

time, 9, 11–17, 19; analogy and, 89, 142, 156, 161, 181–82, 194; archaeonomy and, 142, 160–61, 181, 182, 194; beginning of (*see* beginning; origins); Big Bang and, 142; deep, 46, 47, 90, 149; denial of importance of, 29, 45–47; divine transcendence and, 17; dramatic universe and, 39, 43–44, 107; early Christian understanding of, 155; Einstein and, 12, 18–19, 165–66; escaping from, 29, 35, 152; eternity and, 13, 104, 162; everlasting value of, 198; faith and, 91, 149–50, 151, 152; as fourth dimension, 36; freedom and, 142; fulfillment of, 78, 162; as gift filled with meaning, 94; God and, 11–12, 17, 30, 159, 170; hopeful sense of, 197–98; immensity of, 34, 42, 56, 110–11; indestructible rightness and, 13, 15, 16, 17; intelligibility and, 19, 27; Jesus and, 13, 29–31; local events and, 39; matter's relationship with, 16, 17, 19, 28, 36, 37, 39, 199–200; nature and, 26, 40; in Nicene Creed, 15, 30, 31, 32–33, 197–98; pantheism in, 22; reality of, 16; relativity and, 28, 36, 37, 39–40; resurrection and, 157–58; seeking escape from, 90, 91; stories and, 40; strong, 84–85; subordination to geometry, 73, 152; theology and, 15, 156; unflowing, 18–19; Whitehead on, 165–68. *See also* eternity; immensities, cosmic; narratives
time, irreversible, 15, 18, 95; analogy and, 89; denial of, 152; Einstein and, 16–17, 141; entropy and, 84; evidence of, 36 (*see also* entropy); intelligibility and, 19; in Nicene Creed, 33; science and, 47; stories and, 40
time, passage of, 17, 73, 181–82; evolution and, 26; as illusion, 18, 35, 87; in-

destructible rightness and, 15; meaning in, 67
time, reality of, 2, 166; analogy and, 194; archaeonomy and, 194; evil and, 179, 180–82
timelessness, 12, 35, 47; analogy and, 186, 191–92; ecological responsibility and, 186; Einstein's love of, 73; of geometry, 166; of God, 25; longing for, 156–57; lure of, 2; nostalgia for, 151; perfection associated with, 180. *See also* eternity
transcendence, God's, 47
transformations, physical, 93
Trinity, doctrine of, 13, 31–32
trust, of mind, 115–17, 118, 119, 121–22, 124, 127–28, 147–49
truth, 63, 114, 145–46. *See also* values
Truth-Itself, 119

understanding, 74, 114. *See also* intelligibility; mind/thought
unity, 150. *See also* coherence; intelligibility
universe, 1; age of, 34–35, 42, 81–82; awakening of, 60; beginning in time, 80; collapse of, 83, 84, 156–57, 159–61, 163; as convergent, 109; dramatic meaning of, 43–44; end of, 83, 84, 156–57, 159–61, 163 (*see also* perishing); expansion of, 80–81, 82, 113; finite, 83; fixed, 1, 25, 181; God identified with, 21; ideal, for Einstein, 60; as impersonal, 52; justified by faith, 144; narrative shape of, 95; new understanding of, 9; as pointless, 114, 123; as story, 154; as unfinished. *See also* Big Bang; intelligibility; universe, unfinished

universe, unfinished, 6–7, 20, 27, 47, 58, 59, 67, 68, 200; denial of irreversible time and, 19; not-yet and, 24–25; patience and, 66; relativity and, 7, 38, 59, 154; thought and, 39. *See also* awakening; not-yet
unpredictability, 66

values, 10, 63, 114
variations, accidental, 26. *See also* evolution
Vatican II, 170
virtue: as cosmic development, 171–72; ecological importance of, 185, 192; in fixed universe, 1. *See also* compassion
vitalism, 105–7, 108, 112, 120, 125, 189. *See also* analogy
vocation, 200

waiting, 201; for archaeonomists, 147; faith and, 47, 149, 151; futility of, 25; for return of Jesus, 155 (*see also* hope; resurrection)
Watson, James, 101
Weinberg, Steven, 123
Whitehead, Alfred North, 34, 95, 164, 165–68, 170
Wilson, David Sloan, 117, 119
Wilson, E. O., 188, 189
Wittgenstein, Ludwig, 63
word of God, 11
wrongness, 9; healing and, 180; inability to make sense of, 179; moral, 173–74; natural, 173; nature's inseparability from, 175; sense of, 175, 176 (*see also* compassion)

Yahweh, 10